Core Books in Advanced Mathematics

Newton's Laws and Particle Motion

Core Books in Advanced Mathematics

General Editor: C. PLUMPTON, Moderator in Mathematics,
University of London School Examinations Department;
formerly Reader in Engineering Mathematics,
Queen Mary College, University of London.

Titles available

Newton's Laws and Particle Motion
Mechanics of Groups of Particles
Methods of Trigonometry
Coordinate Geometry and Complex Numbers
Proof
Differentiation
Integration
Vectors
Curve Sketching

Core Books in Advanced Mathematics

Newton's Laws and Particle Motion

Tony Bridgeman
Chief Examiner in Advanced Level Mathematics, University of London School Examinations Department; Senior Lecturer in Applied Mathematics, University of Liverpool.

P. C. Chatwin
Assessor in Advanced Level Mathematics, University of London School Examinations Department; Senior Lecturer in Applied Mathematics, University of Liverpool.

C. Plumpton
Moderator in Mathematics, University of London School Examinations Department; formerly Reader in Engineering Mathematics, Queen Mary College, University of London.

M

Macmillan Education
London and Basingstoke

First published 1984

Published by
Macmillan Education Limited
Houndmills Basingstoke Hampshire RG21 2XS
and London
Associated companies throughout the world

Typeset in Hong Kong by Asco Trade Typesetting Ltd.
Printed in Hong Kong

Contents

Preface

Advanced level mathematics syllabuses are once again undergoing changes in content and approach following the revolution in the early 1960s which led to the unfortunate dichotomy between 'modern' and 'traditional' mathematics. The current trend in syllabuses for Advanced level mathematics now being developed and published by many GCE Boards is towards an integrated approach, taking the best of the topics and approaches of modern and traditional mathematics, in an attempt to create a realistic examination target through syllabuses which are maximal for examining and minimal for teaching. In addition, resulting from a number of initiatives, core syllabuses are being developed for Advanced level mathematics consisting of techniques of pure mathematics as taught in schools and colleges at this level.

The concept of a core can be used in several ways, one of which is mentioned above, namely the idea of a core syllabus to which options such as theoretical mechanics, further pure mathematics and statistics can be added. The books in this series involve a different use of the core idea. They are books on a range of topics, each of which is central to the study of Advanced level mathematics; they form small studies of their own, of topics which together cover the main areas of any single-subject mathematics syllabus at Advanced level.

Particularly at times when economic conditions make the problems of acquiring comprehensive textbooks giving complete syllabus coverage acute, schools and colleges and individual students can collect as many of the core books as they need to supplement books they already have, so that the most recent syllabuses of, for example, the London, Cambridge, AEB and JMB GCE Boards can be covered at minimum expense. Alternatively, of course, the whole set of core books gives complete syllabus coverage of single-subject Advanced level mathematics syllabuses.

The aim of each book is to develop a major topic of the single-subject syllabuses, giving essential book work, worked examples and numerous exercises arising from the authors' vast experience of examining at this level. Thus, as well as using the core books in either of the above ways, they are ideal for supplementing comprehensive textbooks, by providing more examples and exercises, so necessary for the preparation and revision for examinations.

A feature of the books in this series is the large number of worked examples which are regarded as an integral part of the text. It is hoped that these books will be of use not only to schools but also to the large minority of candidates who are studying by themselves. Mechanics, perhaps more so than other branches of mathematics, can be mastered only by working conscientiously and thoughtfully through many worked and unworked examples. This is the only way in which the essential simplicity of the subject can be appreciated.

In order to achieve a degree of uniformity in the size of the books in this series, the Advanced level mechanics topics are covered in two books. In this book the basic laws of mechanics, known as Newton's laws, are considered and applied to particle motion. The second book, *Mechanics of Groups of Particles*, assumes Newton's laws and considers their application to groups of particles, including rigid bodies. The second book also deals with statics.

The treatment in both books assumes a knowledge of vector algebra and of elementary calculus (including the exponential and logarithm functions). These subjects are covered in other volumes in the series. The worked examples and problems have been carefully chosen to be physically important (as distinct from artificial examples devoid of any relevance to the real world). The treatment unashamedly makes appeals to intuition and common sense and the ordering of the topics (sometimes unconventional) has been chosen to stress that all mechanics is based on Newton's laws, not on a series of *ad hoc* techniques. One final feature is the stress on dimensional consistency and the possibility of checking that this allows.

While the notation used is generally self-explanatory, it should be noted that there are two distinct ways of specifying the inverse trigonometric functions. Thus $\tan^{-1} x$ and $\arctan x$ are both used to denote the angle θ between $-\pi/2$ and $\pi/2$ such that $\tan\theta = x$. Also, the symbol $\approx$ means 'approximately equals'.

Tony Bridgeman
P. C. Chatwin
C. Plumpton

1 Introduction

1.1 The scope of mechanics

Mechanics is the study of how bodies respond when forces are applied to them. Both experiment and theory can be used in this study; this book is primarily concerned with theory. Because the theory uses mathematical tools, mechanics studied in this way is a branch of *applied mathematics*, the general name given to work in which mathematics is used to explain and predict phenomena occurring in the real world. The importance and value of applied mathematics in general, and theoretical mechanics in particular, lies not in the elegance of its mathematical structure but in its success in aiding the understanding of real phenomena. It follows that the theoretical study of mechanics has no value unless its results can be compared with actual events.

It is a matter of common experience that bodies may, or may not, move when forces are applied to them. Thus, for example, a book remains at rest when placed on a table even though the table is supporting the book by exerting a force upon it; on the other hand, a football moves when a force is exerted on it (by a kick). The branch of mechanics dealing with cases not involving motion is called *statics*, and the branch concerned with situations in which motion takes place is called *dynamics*. This book deals with the fundamental principles of dynamics.

However, since this book is an introduction, only simple bodies will be considered. It turns out in fact that in many circumstances the size of a moving body is unimportant. For example, the size of the Earth hardly affects the way in which it moves round the Sun (but its size obviously does affect the way we human beings move on its surface!). The term *particle* will be used to denote a body whose size can be ignored in a particular situation. In some cases when the size of the body cannot be neglected, the body may legitimately be regarded as a collection of particles rigidly joined together; it is then termed a *rigid body*. In this book we consider only the mechanics of particles. However, many of the basic concepts can be applied to more complicated situations.

1.2 Mass, time, length and angle

Mathematics can be used to study mechanics only because the important properties of bodies and their motion can be quantified; that is they can be measured and expressed in terms of numbers.

From the point of view of mechanics, the most fundamental property possessed by a body is its *mass*. Roughly speaking, the mass of a body is a measure of the quantity of matter that it contains. More precisely, two bodies are said to have the same mass if they exactly balance one another when placed on opposite pans of an accurate pair of scales. This idea is of course used throughout commerce, and a shopkeeper measures the mass of goods to be sold by balancing them against some of his standard masses.

The same idea is used in science. A certain body, made of a mixture of platinum and iridium, and kept in Sèvres near Paris, was adopted in 1901 as the standard, and was defined to have a mass of 1 *kilogram* (abbreviated to 1 kg).

Another body has a mass of 1 kg if it exactly balances the standard on an accurate pair of scales. Using this principle, many bodies of mass of 1 kg were constructed and distributed throughout the world. Furthermore, by careful workmanship, one of these bodies of mass 1 kg was divided into two smaller bodies which balanced one another on a pair of scales. Each of these smaller bodies has the same mass, namely 0·5 kg. By repeated extension of this process, many different standard masses were constructed ranging from very small masses to very large masses. The mass of a given body can then be determined, in a laboratory or a shop, by placing it on one pan of a pair of accurate scales and by putting suitable standard masses on the other pan until an exact balance is obtained. Trial and error is, of course, a normal part of this process.

In discussing the division of a body of mass 1 kg into two smaller bodies each of mass 0·5 kg, it was assumed that the total mass was not changed by the process of division. The assumption that mass cannot be increased or decreased by any process was, for many centuries, a fundamental belief of mechanics known as the *principle of conservation of mass*. This is consistent with the results of a wide range of experiments and, until the development of relativity, was believed to be true in all circumstances. We now know that the principle needs to be modified when, but only when, the bodies involved are moving at speeds close to that of light. Since such motions will not be considered in this book, the principle of conservation of mass can be assumed.

As explained earlier, a particle is a body whose size can be neglected when studying its motion. However, experiments show that the mass of a body does affect its motion and cannot be ignored. It makes sense therefore to refer for example to a particle of mass 1·34 kg, or, more generally, to a particle of mass m, where m stands for a number of kilograms (not just a number).

Other quantities have to be measured in order to describe how a body moves. One obvious example is *time*. Historically time was measured in terms of astronomical observations so that, for example, a year was the period in which the Earth moved once around the Sun, and a day was the time taken for the Earth to rotate once about its own axis. It was found that such definitions were unsatisfactory for precise work because, for example, the duration of a year varies due to influences on the Earth of bodies other than the Sun. More satisfactory standards were sought, and since 1967 time has been defined internationally in terms of the radiation from the caesium-133 atom; the standard measure of time is 1 *second* (abbreviated to 1 s). For practical purposes this definition of a second is the same as that used in everyday life; thus, for example, 86 400 s is 1 day.

Motion is described in terms of distance travelled as well as time, so that *length* also has to be measured. The standard length, internationally agreed, is 1 *metre* (abbreviated to 1 m). Originally the metre was defined in terms of a distance between two points on the Earth's surface, and later as the distance between two marks on a piece of metal, which, like the standard kilogram, was also kept near Paris. Since 1960 this definition has also been superseded by one in terms of atomic radiation, in this case the krypton-86 atom.

Another basic quantity is *angle*, needed for example to specify the direction of motion of a body. There are several ways of measuring angle which all follow the same principle of expressing the amount of turn between two directions as a fraction of one complete turn. The two most familiar systems are those using *degrees* (°), in which one complete revolution is 360°, and *radians* (rad), in which one complete revolution is 2π rad. For algebraic convenience the system using radians will usually be employed in this book, but, since an angle is a pure number, the abbreviation 'rad' will often be dropped. So, for example, an angle of $(\pi/6)$ will mean an angle of $(\pi/6)$ rad, which is 30°; and $\sin(\pi/4)$ is $\sin(\pi/4 \text{ rad}) = \sin(45°) = \sqrt{2}/2$.

1.3 Units and dimensions

The system in which the basic units of length, mass and time are the metre, kilogram and second, respectively, is the International System of Units (SI). In scientific work this is now gradually replacing other systems such as the CGS system, in which the basic units are the centimetre (cm), gram (g) and second, and the FPS system, in which the basic units are the foot (ft), pound (lb) and second. There are, of course, relations between the basic units in different systems, such as $1\text{ cm} = 10^{-2}\text{ m}$, $1\text{ ft} = 30{\cdot}48 \times 10^{-2}\text{ m}$, $1\text{ g} = 10^{-3}\text{ kg}$, $1\text{ lb} \approx 0{\cdot}4536\text{ kg}$, and these are used when it is necessary to convert from one system of units to another. The following example illustrates this process.

Example 1 A mathematical model of the Earth is that it is a sphere of approximate radius $6{\cdot}37 \times 10^6$ m and approximate mass $5{\cdot}98 \times 10^{24}$ kg. Estimate its radius in ft and its mass in lb. Find also its average density in (a) kg m^{-3}, (b) lb ft^{-3}.

Since $1\text{ ft} = 30{\cdot}48 \times 10^{-2}\text{ m}$, $1\text{ m} = (10^2/30{\cdot}48)\text{ ft}$.

$$\therefore \text{ Radius of Earth} \approx 6{\cdot}37 \times 10^6\text{ m} = (6{\cdot}37 \times 10^8/30{\cdot}48)\text{ ft}$$
$$\approx 2{\cdot}09 \times 10^7\text{ ft}.$$

Similarly $1\text{ lb} \approx 0{\cdot}4536\text{ kg} \Rightarrow 1\text{ kg} \approx (1/0{\cdot}4536)\text{ lb}$.

$$\therefore \text{ Mass of Earth} \approx 5{\cdot}98 \times 10^{24}\text{ kg} \approx (5{\cdot}98 \times 10^{24}/0{\cdot}4536)\text{ lb}$$
$$\approx 1{\cdot}32 \times 10^{25}\text{ lb}.$$

The average density of a body is its mass divided by its volume. Assuming the Earth is a sphere, its volume is

$$\frac{4\pi}{3}(6{\cdot}37 \times 10^6)^3 \text{ m}^3 \approx \frac{4\pi}{3}(2{\cdot}09 \times 10^7)^3 \text{ ft}^3,$$

that is $1{\cdot}08 \times 10^{21}$ m^3 or $3{\cdot}82 \times 10^{22}$ ft^3. Hence its average density is

(a) $(5{\cdot}98 \times 10^{24}/1{\cdot}08 \times 10^{21})$ kg m$^{-3} \approx 5{\cdot}54 \times 10^3$ kg m^{-3}, or

(b) $(1{\cdot}32 \times 10^{25}/3{\cdot}82 \times 10^{22})$ lb ft$^{-3} \approx 3{\cdot}46 \times 10^2$ lb ft^{-3}.

In this book letters will often be used to stand for properties of bodies and their motion. The example of a particle of mass m has already been mentioned. Similarly, it will be necessary and convenient to refer to a time t or T, and to a length x or a. Other letters will often be used to denote other physical quantities. One of the purposes of theoretical mechanics is to obtain relationships between different quantities, such as a relationship between the distance x travelled by a particle and the time t it has been travelling.

These relationships will usually be in the form of mathematical formulae, but because they are relationships between quantities with units, not all mathematical formulae are acceptable. This is a consequence of the fact that when a letter is used to represent a physical quantity it stands for a number of units and not just a number. For example, if the letter R denotes the radius of the Earth then, by Example 1, each of the equations $R = 6{\cdot}37 \times 10^6$ m and $R = 2{\cdot}09 \times 10^7$ ft is correct (to 3 significant figures) and therefore so is the equation $6{\cdot}37 \times 10^6$ m $= 2{\cdot}09 \times 10^7$ ft. Without the units m and ft, however, the equation is nonsense.

Obviously mathematical formulae relating different quantities must not allow nonsensical results. This means they must give correct results whatever units are used to measure the different quantities.

Suppose a car is travelling along a road at a steady speed V, and let s denote the distance it travels in a time t. Then s and t are related by the equation

$$s = Vt. \tag{1.1}$$

Consider a case when V is 30·00 mph. In $\frac{1}{2}$ h (half an hour) the distance travelled is $(30{\cdot}00 \times \frac{1}{2})$ miles $= 15{\cdot}00$ miles, using (1.1). Suppose further that during this period the car is followed at a constant distance by a French car. According to the speedometer of this car V is 48·28 kph, where kph (often written km h^{-1}) means kilometres (km) per hour and 1 km $=$ 1000 m. In $\frac{1}{2}$ h the distance travelled is $(48{\cdot}28 \times \frac{1}{2})$ km $= 24{\cdot}14$ km. Since 1 mile $= 1{\cdot}609$ km, it follows that 15·00 miles $= 24{\cdot}14$ km, so that the two results are the same. It can be checked that (1.1) is also correct if other units are used (e.g. V measured in ft s^{-1}, s in ft, and t in s).

On the other hand, suppose the formula

$$s = 30\,t \tag{1.2}$$

is used instead of (1.1). This gives the correct numerical value of s for the first driver whose units of distance and time are miles and hours respectively, but it gives the *wrong* values when other units are used. Thus (1.2) is not a satisfactory alternative to (1.1).

The reason why (1.2) must be rejected can also be understood by considering the concept of a *dimension*. All properties considered in theoretical mechanics have units formed from the three basic units of mass, length and time. Thus the unit of speed is one unit of length per unit of time, and the unit of density is one unit of mass per cube of the unit of length. Since these statements are true irrespective of the particular system of units used, the term dimension is used here instead of unit. Thus, using the symbols M, L, T to denote mass, length and time respectively, the dimension of speed is LT^{-1} and that of density is ML^{-3}. The left-hand side of (1.1) is a length and so has dimension L, whereas the right-hand side is the product of a speed and a time and so has dimension $LT^{-1}.T = L$. Equation (1.1) is therefore *dimensionally consistent*. On the other hand, the left-hand side of (1.2) has the dimension L whereas the right-hand side has dimension T, and so (1.2) is dimensionally inconsistent.

This example illustrates that all equations in theoretical mechanics must be dimensionally consistent.

Example 2 A particle is oscillating along a straight line and the time taken for one complete oscillation (the *period* of the motion) is t_0. Let x denote the displacement of the particle from a fixed point O on the line at a time t after passing through O, and suppose X is the maximum value of x. Show that the equation

$$x = X \sin(2\pi t/t_0) \tag{1.3}$$

is dimensionally consistent.

The dimension of $(2\pi t/t_0)$ is zero; that is $(2\pi t/t_0)$ is a pure number. This is because 2π is a pure number ($\approx 6{\cdot}283$), as is (t/t_0) which is the ratio of two quantities each with the dimension T and is therefore independent of the units used for measuring time. Hence $\sin(2\pi t/t_0)$ is a pure number whose value is also independent of the units used for measuring time. The quantities x and X each have the dimension L, so (1.3) is dimensionally consistent.

1.4 Basic mathematical techniques

Certain basic mathematical skills are needed to develop the subject of theoretical mechanics. Among these are calculus (including the differentiation and integration of simple polynomials, trigonometric, logarithmic and exponential functions), trigonometry (including the meanings of quantities such as $\sin\theta$, $\cos\theta$ and $\tan\theta$ and formulae such as $\cos(\theta+\phi) \equiv \cos\theta\cos\phi - \sin\theta\sin\phi$ and $\sin 2\theta = 2\sin\theta\cos\theta$) and vector algebra (as described in a companion book to this one entitled *Vectors*). In the remainder of this book it will be assumed that the reader has mastered these skills.

Exercise 1

1 Express the angle of 11·25 rad in °, giving your answer to the nearest °. Also determine the sine, cosine and tangent of this angle correct to two decimal places.

2 A unit of distance used in astronomy is the *light-year*, defined to be the distance travelled by light in one year of 365 days. Given that the speed of light is approximately $3{\cdot}00 \times 10^8$ ms^{-1}, estimate the distance of 1 light-year in (a) m; (b) km; (c) miles. Give your answers to three significant figures.

3 A ship is sailing at 20 knots, where 1 knot = 1 nautical mile per hour and 1 nautical mile = 6080 ft. Estimate the speed of the ship in ms^{-1}.

4 A ball is thrown vertically into the air with starting speed u. After an interval of time t it has travelled a distance s and its speed is v. Show that the following formulae are dimensionally consistent provided the constant g has dimension LT^{-2}:

$$v^2 = u^2 - 2gs; \quad v = u - gt.$$

Also, determine which, if any, of the following three formulae are dimensionally consistent:

$$s = ut - \tfrac{1}{2}gt^2; \quad s^2 = g^2t^4 + (u^2/2g); \quad v = \exp(gt/u).$$

5 A small sphere of radius a is made of uniform material of density ρ. It is released from rest in a liquid of density ρ' and falls vertically. At a time t after release its speed is v, where

$$v = \frac{2}{9}\left(\frac{\rho - \rho'}{\rho'}\right)\left(\frac{ga^2}{\mathrm{v}}\right)\left[1 - \exp\left(-\frac{9\mathrm{v}t\rho'}{2a^2\rho}\right)\right],$$

and g and v are positive constants. Show that this equation is dimensionally consistent provided g has dimension LT^{-2} and v has a specific dimension to be determined.

Show also that

$$\frac{\mathrm{d}v}{\mathrm{d}t} + kv = l,$$

where k and l are constants. What are the dimensions of k and l?

Draw a suitable sketch graph which shows how v varies with t, indicating particularly how v behaves for large values of t.

2 Kinematics

2.1 Definition

Before we can use mathematics to predict how bodies move, we must first learn how to describe motion itself. The study of motion without reference to the forces causing the motion is called *kinematics* and it is the subject of this chapter.

2.2 Motion in a straight line

It is convenient to begin with an important special case, namely that in which the body moves in a straight line; for example, a car moving on a straight road, or a stone falling vertically from the top of a cliff.

By carefully observing a motion of this type we can, in principle, plot points on a graph showing how the distance travelled by the body varies with time. Of course distance and time both have to be measured from convenient starting points. Possibilities for the case of a car are to measure distance from a garage or signpost. Examples of graphs showing how distance varies with time in two different cases are given in Fig. 2.1 and Fig. 2.2.

As explained in Chapter 1, it is usually convenient to introduce letters to stand for time and distance. Here we shall generally use t for time and x for distance, but other symbols would do equally well, such as s for distance, as in (1.1). There will be one value of x for each value of t; in other words x is a *function* of t. The work in §1.3 shows that this functional relationship between x and t must be dimensionally consistent, as in (1.3), p. 5. You should check that the curve in Fig. 2.1 is that of $x = \frac{1}{2}gt^2$, where g is a constant of dimension LT^{-2} whose value is 10 ms^{-2}, and you should confirm that this relationship is dimensionally consistent. In general the relationship between x and t is not expressible as a simple formula. One such example is illustrated in Fig. 2.2 which describes the motion of a car on a road; it is a valuable exercise to suggest possible reasons for the shape of this graph.

Only the position of the body at time t has so far been considered and it is measured by the *displacement* x. Just as important is the *speed* of the body at time t. We shall usually denote speed by u or v. By definition the speed is the result of *differentiating* x with respect to t. Hence, using v to denote speed,

$$v = \frac{dx}{dt}. \qquad (2.1)$$

Equivalently, v is the *derivative* or *rate of change* of x with respect to t.

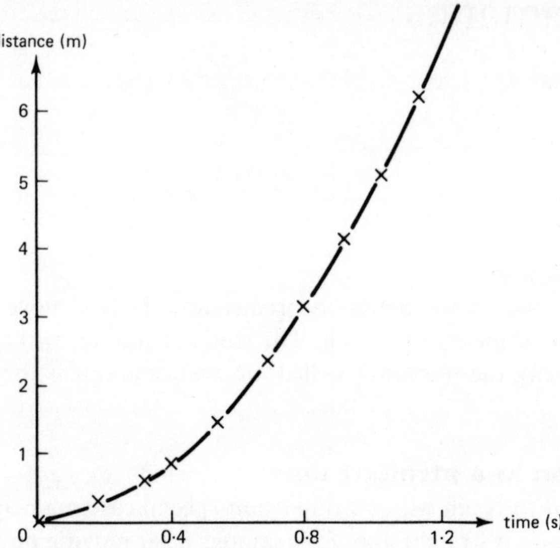

Fig. 2.1 Graph showing how the distance fallen by a stone varies with time. The crosses indicate measured points.

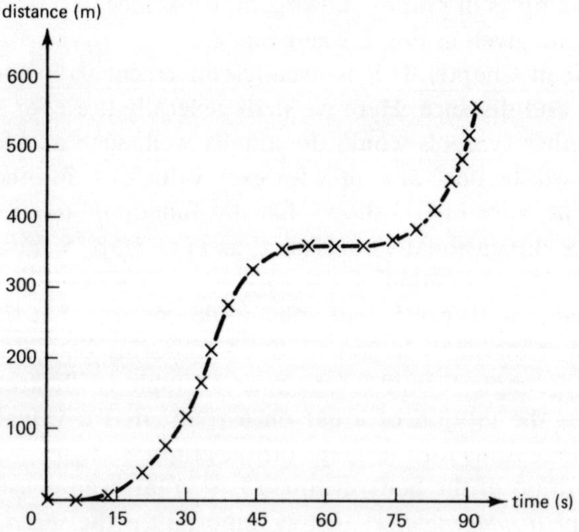

Fig. 2.2 Graph showing how the distance travelled by a car along a road varies with time. The crosses indicate measured points.

In general v varies with t, as does x itself. So we can also differentiate v with respect to t. The result is an important quantity called the *acceleration*. When a separate letter is needed for the acceleration we shall usually employ a, although some authors use f.

Hence

$$a = \frac{dv}{dt} = \frac{d^2x}{dt^2}. \tag{2.2}$$

We have seen that Fig. 2.1 is the curve of $x = \frac{1}{2}gt^2$, where g is a constant equal to 10 ms^{-2}. It follows from (2.1) that $v = gt$, and from (2.2) that $a = g$. Hence Fig. 2.1 illustrates a motion with a constant acceleration of 10 ms^{-2}, and we shall reconsider this type of motion later.

Example 1 A particle moves along a straight line so that its displacement x from a fixed point at time t is given by

$$x = X \sin \omega t, \tag{2.3}$$

where X and ω are positive constants. Find the speed v and the acceleration a at time t.

$$(2.1) \quad \Rightarrow \quad v = \frac{d}{dt}[X \sin \omega t] = \omega X \cos \omega t.$$

$$(2.2) \quad \Rightarrow \quad a = \frac{d}{dt}[\omega X \cos \omega t] = -\omega^2 X \sin \omega t.$$

Note that (2.3) is the same as (1.3), p. 5, provided that $\omega = 2\pi/t_0$ and is therefore dimensionally consistent provided the dimension of ω is T^{-1}. This example illustrates that a usually varies with t and can take positive and negative values, as also can v and x.

It is sometimes useful to express v in terms of x rather than t. For example when $x = \frac{1}{2}gt^2$, then $v = gt$ becomes $v = (2gx)^{1/2}$. When v is expressed in terms of x we cannot directly use (2.2) to find a. But it is easy to derive an alternative formula. By the *chain rule* of calculus (sometimes called the *function of a function rule*) it follows from (2.2) that

$$a = \frac{dv}{dt} = \frac{dv}{dx} \times \frac{dx}{dt}.$$

Since $v = \dfrac{dx}{dt}$, it follows immediately that

$$a = v\frac{dv}{dx} = \frac{d}{dx}(\tfrac{1}{2}v^2). \tag{2.4}$$

When $v = (2gx)^{1/2}$, $v^2 = 2gx$ and (2.4) gives $a = g$, as obtained earlier by direct use of (2.2).

In theoretical mechanics we often use a shorthand notation to denote derivatives with respect to time t. We put one dot over any quantity (such as x or v) each time it is differentiated with respect to t. Using this notation, (2.1) and (2.2) can be rewritten

$$v = \dot{x}, \qquad a = \dot{v} = \ddot{x}. \tag{2.5}$$

Similarly one dash is often used to denote a differentiation with respect to x, so that (2.4) becomes

$$a = vv' = (\tfrac{1}{2}v^2)'. \tag{2.6}$$

We have obtained a by differentiating x twice with respect to t, and we could define new quantities by differentiating x more than twice. However it turns out that this is not useful in mechanics. As will be seen later, the basic physical laws give an expression for a in terms of the *forces* acting on a body. The fundamental problem of mechanics, and therefore the main theme of this book, is to use this expression for a to determine v, and hence x, in terms of t. This process is the reverse of what we have so far done, and therefore involves *integration* rather than differentiation.

Consider a graph of x against t such as Fig. 2.1 or Fig. 2.2, p. 8. The graph of v against t is obtained by plotting the *gradient* of x in this graph against t. The graph of a against t is obtained in an analogous manner from the graph of v against t. Since $a = \dfrac{\mathrm{d}v}{\mathrm{d}t}$, we know from the basic principles of integration that the change in v between any two times t_1 and t_2, where $t_2 > t_1$, is the area under the graph of a against t between t_1 and t_2, that is

$$\Big[v(t)\Big]_{t_1}^{t_2} = v(t_2) - v(t_1) = \int_{t_1}^{t_2} a(t)\,\mathrm{d}t. \tag{2.7}$$

This is illustrated in Fig. 2.3. Similarly, since $v = \dfrac{\mathrm{d}x}{\mathrm{d}t}$, we have

$$\Big[x(t)\Big]_{t_1}^{t_2} = x(t_2) - x(t_1) = \int_{t_1}^{t_2} v(t)\,\mathrm{d}t. \tag{2.8}$$

In (2.7) and (2.8), t_1 and t_2 can be *any* values of t.

Example 2 A body is moving in a straight line with constant acceleration g. Given that at $t = t_0$ the speed is V and the displacement from a fixed point in the straight line is X, determine the speed and displacement for all values of t.

$$(2.7) \quad \Rightarrow \quad v(t_2) - v(t_1) = \int_{t_1}^{t_2} g\,\mathrm{d}t = g(t_2 - t_1).$$

Since we know what happens at time t_0, we put $t_1 = t_0$, $v(t_1) = V$ to obtain $v(t_2) - V = g(t_2 - t_0)$. Since t_2 can be any value of t, we put $t_2 = t$ and rearrange to find the speed v. The result is

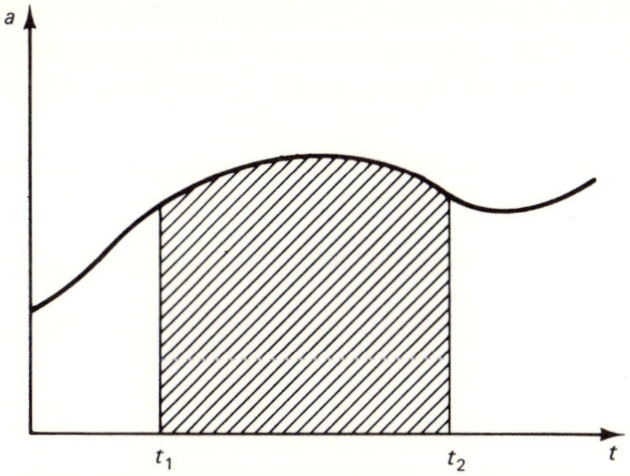

Fig. 2.3 Sketch graph illustrating equation (2.7). The shaded area is the change in the body's speed between t_1 and t_2.

$$v(t) = V + g(t - t_0). \tag{2.9}$$

$$(2.8) \quad \Rightarrow \quad x(t_2) - X = \int_{t_0}^{t_2} [V + g(t - t_0)]\,\mathrm{d}t$$
$$= V(t_2 - t_0) + \tfrac{1}{2}g(t_2 - t_0)^2.$$

Here we have put $t_1 = t_0$, $x(t_1) = X$ immediately. Once more we argue that, since t_2 is any value of t, we can put $t_2 = t$. Rearrangement gives

$$x(t) = X + V(t - t_0) + \tfrac{1}{2}g(t - t_0)^2. \tag{2.10}$$

The result (2.10) will be important later. Here we indicate how it can be used in one example.

Example 3 A car moving along a straight road with constant speed V passes a point O at the same moment as a second car starts to accelerate with constant acceleration f from rest at O. Find the time after leaving O when they are again level, and show that the distance covered by each car is then $2V^2/f$.

Let x_1 be the distance from O of the first car at time t after leaving O. We can use (2.10) with $X = g = t_0 = 0$ to obtain

$$x_1 = Vt.$$

Let x_2 be the distance from O of the second car at time t after leaving O. We can again use (2.10) with $X = V = t_0 = 0$ and $g = f$ to obtain

$$x_2 = \tfrac{1}{2}ft^2.$$

They are level when

$$x_1 = x_2 \quad \Rightarrow \quad Vt = \tfrac{1}{2}ft^2$$
$$\Rightarrow \quad t = 0 \text{ or}$$
$$t = 2V/f.$$

They leave O when $t = 0$, so they are again level when $t = 2V/f$.

$$t = 2V/f \quad \Rightarrow \quad x_1 = x_2 = 2V^2/f.$$

Sometimes the laws of mechanics do not give a as a function of t but as a function of (i) x or (ii) v. In case (i) we can use (2.4), p. 9, and integrate to obtain

$$\Big[v^2\Big]_{x_1}^{x_2} = v^2(x_2) - v^2(x_1) = 2\int_{x_1}^{x_2} a(x)\,\mathrm{d}x. \tag{2.11}$$

Example 4 An artificial satellite is fired vertically from the Earth's surface. When it is at a distance x from the centre of the Earth its acceleration is $-GM/x^2$, where G is Newton's constant of gravitation and M is the mass of the Earth. Find the speed at which the satellite should be fired for it to escape from the Earth. Use the following approximate numerical values:

$$G = 6{\cdot}67 \times 10^{-11}\ \mathrm{m^3\,kg^{-1}s^{-2}},\ M = 5{\cdot}98 \times 10^{24}\ \mathrm{kg},$$

$$\text{Earth's radius} = 6{\cdot}37 \times 10^6\ \mathrm{m}.$$

The minus sign in the expression for the acceleration means that the speed of the satellite decreases as x increases. Denote the required speed by V and the Earth's radius by R. We use $a(x) = -GM/x^2$, $x_1 = R$, $v(x_1) = V$ in (2.11) to obtain

$$v^2(x_2) - V^2 = 2\int_R^{x_2} -\frac{GM}{x^2}\,\mathrm{d}x = -2GM\int_R^{x_2}\frac{1}{x^2}\,\mathrm{d}x$$

$$= -2GM\left[-\frac{1}{x}\right]_R^{x_2} = -2GM\left(\frac{1}{R} - \frac{1}{x_2}\right).$$

Hence, writing $x_2 = x$, $v(x_2) = v$, we find

$$v^2 = \left(V^2 - \frac{2GM}{R}\right) + \frac{2GM}{x}. \tag{2.12}$$

From (2.12) it follows that v^2 (and so v) decreases as x increases. There are two possibilities. Either there is a value of x for which $v = 0$ (in which case the satellite reaches this value and then falls back to the Earth), or there is no such value (in which case the satellite continually recedes from the Earth, eventually escaping its influence altogether). We are interested only in the second possibility which occurs when the right-hand side of (2.12) is positive for all x. Hence we must have $V^2 \geqslant 2GM/R$. Substituting the given numerical values yields

$$V^2 \geqslant \left(\frac{2 \times 6{\cdot}67 \times 5{\cdot}98}{6{\cdot}37}\right) \times 10^7\ \mathrm{m^2s^{-2}},$$

$$\Rightarrow \quad V^2 \geqslant 12{\cdot}52 \times 10^7\ \mathrm{m^2s^{-2}}$$

$$\Rightarrow \quad V \geqslant 11{\cdot}2 \times 10^3\ \mathrm{ms^{-1}}.$$

(This least value of V of $11{\cdot}2 \times 10^3$ ms^{-1} is known as the escape velocity for the Earth.)

Example 4 illustrates the sensible rule that when a numerical result is required it is better to substitute the numbers only after all the algebra has been done. Thus, for example, we used R, not $6{\cdot}37 \times 10^6$ m, until we had an expression for the answer. There are four advantages in this:

(i) careless mistakes are less likely to occur;
(ii) working algebraically means that checks of dimensional consistency can be made (e.g., check that (2.12) is dimensionally consistent);
(iii) the algebra is exact whereas the numbers are only approximate;
(iv) there is less writing to do (if you disbelieve this, rework Example 4 substituting numbers from the start!).

When (2.11) is used, as in Example 4, to obtain v in terms of x, it is not always easy to go further and obtain x in terms of t. Consider (2.12) for example, in the case when $V^2 > 2GM/R$. Writing

$$W^2 = (V^2 - 2GM/R) \quad \text{and} \quad X = 2GMR/(RV^2 - 2GM)$$

enables (2.12) to be rewritten

$$v^2 = W^2[1 + (X/x)].$$

Since $v = \dfrac{dx}{dt}$ and since $v > 0$ (because the satellite is moving so that x increases with t), we can take the positive square root to obtain

$$\frac{dx}{dt} = W\left(1 + \frac{X}{x}\right)^{1/2}. \qquad (2.13)$$

It is possible to find x in terms of t from (2.13), and this is the subject of question 5 in Exercise 2 at the end of this chapter.

As mentioned above, the laws of mechanics sometimes give the acceleration as a function of v. Example 5 is typical of such cases.

Example 5 A body falling through thick oil has an acceleration $(g - kv)$ downwards, where g and k are positive constants. The body is released from rest at $t = 0$. Find v in terms of t, and discuss the motion when kt is large.

$$(2.2) \quad \Rightarrow \quad \frac{dv}{dt} = (g - kv)$$

$$\Rightarrow \quad \frac{1}{(g - kv)}\frac{dv}{dt} = 1.$$

The left-hand side is the derivative with respect to v of $-\dfrac{1}{k}\ln(g - kv)$ so that

$$\frac{d}{dt}\left[-\frac{1}{k}\ln(g-kv)\right]=1$$

$$\Rightarrow \quad \frac{d}{dt}[\ln(g-kv)]=\frac{d}{dt}(-kt).$$

If two functions have the same derivative they differ by a constant. Writing C for this constant gives

$$\ln(g-kv)=C-kt$$

$$\Rightarrow \quad (g-kv)=e^{(C-kt)}=e^{C}e^{-kt}.$$

$$v=0 \text{ at } t=0$$

$$\Rightarrow \quad g=e^{C}$$

$$\Rightarrow \quad g-kv=ge^{-kt}$$

$$\Rightarrow \quad v=\left(\frac{g}{k}\right)[1-e^{-kt}].$$

This is the required result. As kt increases, e^{-kt} decreases, eventually to 0. Hence v approaches the *limiting*, or *terminal*, *speed* (g/k).

In Examples 2, 4 and 5 we had to find v in terms of t or x starting with an exact algebraic formula for the acceleration. In many practical cases, such as the one illustrated in Fig. 2.2, there is no simple formula applicable. Often the only information available will be measurements of the acceleration or the speed at some values of x or t, and it is not then possible to determine exactly the integrals on the right-hand sides of (2.7), (2.8) and (2.11). However, provided that a sufficient number of measurements are given, an approximate graph can be drawn and the integral can be estimated numerically by one of several methods. The simplest such technique is to count the squares under the graph; frequently this is the only available method and it is illustrated in question 6 in Exercise 2 (p. 27). In some cases, however, measurements may have been made at values of x or t which are equally spaced, and then other approximate methods can be used. The simplest of these is the *trapezium rule*, the use of which is illustrated in Example 6.

Example 6 The table shows the speeds of a bicycle measured at intervals of 1 s. Use the trapezium rule to estimate the distance travelled between $t = 1$ s and $t = 15$ s.

t(s)	1	2	3	4	5	6	7	8	9	10	11	12	13	14	15
v(ms^{-1})	0·6	1·0	1·5	2·5	3·5	4·6	5·4	6·2	7·1	8·2	9·2	9·9	10·5	10·9	11·1

The measurements are plotted in Fig. 2.4. There is not enough information to

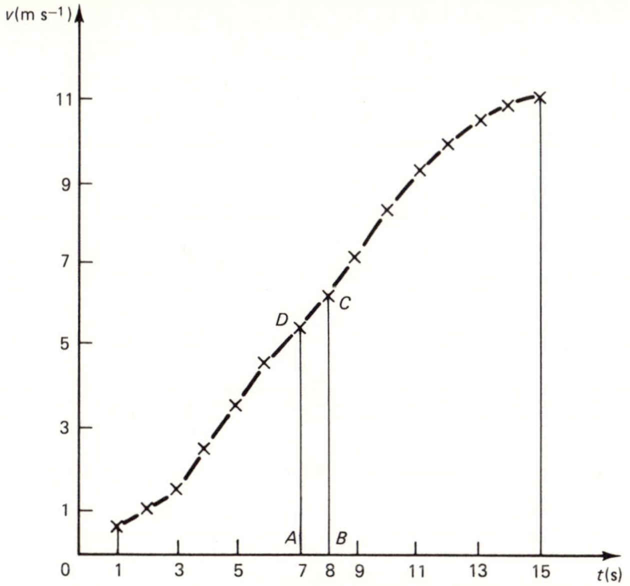

Fig. 2.4 Graph illustrating Example 6.

draw the actual curve between $t = 1$ s and $t = 15$ s, yet, by (2.8), the distance required is the area under this curve between these two values of t. The trapezium rule is based on the assumption that this area is approximately equal to the sum of the areas of trapezia such as $ABCD$. The area of the trapezium $ABCD$ is an approximation to the distance travelled between $t = 7$ s and $t = 8$ s. (The area of a trapezium is one half of the sum of its parallel sides multiplied by the distance between them.) Altogether the required area is approximated by the sum of the areas of the 14 trapezia; so the distance travelled is estimated to be

$$[\tfrac{1}{2}(0{\cdot}6 + 1{\cdot}0) \times 1] + [\tfrac{1}{2}(1{\cdot}0 + 1{\cdot}5) \times 1] + \ldots + [\tfrac{1}{2}(10{\cdot}9 + 11{\cdot}1) \times 1]$$

$$= \tfrac{1}{2} \times [0{\cdot}6 + 2(1{\cdot}0 + 1{\cdot}5 + \ldots + 10{\cdot}9) + 11{\cdot}1] = 86{\cdot}35 \text{ m}.$$

As with all methods of this type the answer is subject to two kinds of error. One arises from the inevitable experimental errors and the other from replacing the actual curve by a series of straight lines. These errors can be estimated but this important subject is beyond the scope of this book. However, without such error analysis it is wise to be cautious about our result; so we say that the distance travelled is about 86·4 m.

The trapezium rule can be applied to most integrals, such as those in (2.7) and (2.11), provided it appears from the measurements that the actual curve can be well approximated by a series of straight lines.

2.3 Motion in space and its description using vectors

Having studied motion along a straight line, we now consider how to describe the most general possible motion of a body, namely motion along a curved path in real three-dimensional space. This can be done in several ways, but the best method uses *vector* notation.

The development of vector concepts and notation is fully described in many books, including *Vectors*, another book in this series. In this section we shall summarize the principal results needed in theoretical mechanics.

We choose a reference point called the *origin* denoted by O. Consider any point P in space. The *directed line segment* $\overrightarrow{OP}$ specifies a vector called the *position vector* of P relative to O. Let $\mathbf{r}$ denote this vector. Now suppose that P is moving so that $\mathbf{r}$ varies with time t. For each value of t there is a unique value of $\mathbf{r}$. In other words $\mathbf{r}$ is a function of t and we write $\mathbf{r}(t)$ instead of $\mathbf{r}$ when we wish to emphasize the dependence on t.

A possible path of P is illustrated schematically in Fig. 2.5 (but it should be realized that the path does not have to lie in a plane). Although the path is curved, the part of it traversed between two times t and $t + \delta t$ can be approximated to a straight line segment, if δt is small, with an accuracy that increases as δt decreases. This point is illustrated in Fig. 2.5. The directed line segment $\overrightarrow{QR}$ specifies the vector $\boldsymbol{\delta}\mathbf{r}$ where

$$\boldsymbol{\delta}\mathbf{r} = \mathbf{r}(t + \delta t) - \mathbf{r}(t). \tag{2.14}$$

The length of $\boldsymbol{\delta}\mathbf{r}$, written $|\boldsymbol{\delta}\mathbf{r}|$, is the small distance travelled by P between times t and $t + \delta t$. Hence, by (2.1), p. 7, as δt approaches zero, $|\boldsymbol{\delta}\mathbf{r}|/\delta t$ approaches the magnitude of the speed of P at time t. But for a particle moving

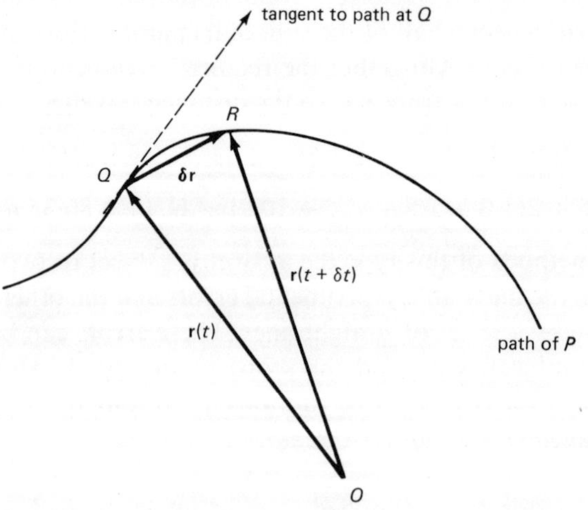

Fig. 2.5 Sketch illustrating the definition of velocity. The points Q and R are the positions of P at times t and $t + \delta t$.

in space it is important to keep track of the *direction* of motion as well as the speed. Since $\boldsymbol{\delta}\mathbf{r}$, defined by (2.14), is in this direction, so also is $\boldsymbol{\delta}\mathbf{r}/\delta t$. The natural generalization of the concept of the speed of a particle moving along a straight line is therefore to consider the limit of $\boldsymbol{\delta}\mathbf{r}/\delta t$ as δt approaches zero. This limit is defined to be the *velocity* of P at time t and is denoted by $\mathbf{v}$, or by $\frac{\mathrm{d}\mathbf{r}}{\mathrm{d}t}$, or by $\dot{\mathbf{r}}$, using the convention defined in (2.5). Thus

$$\mathbf{v} = \frac{\mathrm{d}\mathbf{r}}{\mathrm{d}t} = \dot{\mathbf{r}} = \lim_{\delta t \to 0}\left(\frac{\boldsymbol{\delta}\mathbf{r}}{\delta t}\right) = \lim_{\delta t \to 0}\left(\frac{\mathbf{r}(t+\delta t) - \mathbf{r}(t)}{\delta t}\right). \tag{2.15}$$

It is clear from Fig. 2.5 that the direction of $\mathbf{v}$ is along the *tangent* to the path of P. The magnitude of $\mathbf{v}$, denoted by $|\mathbf{v}|$, is the limit of $|\boldsymbol{\delta}\mathbf{r}|/\delta t$ and so, as we have already noted, is the magnitude of the speed of P, where speed is defined in (2.1). However, it is conventional when using vectors to describe motion in three-dimensional space, to define $|\mathbf{v}|$ as the *speed* of P, so that the speed of a particle is now an essentially positive quantity. This contradicts the use of the term speed in §2.2, which could be positive or negative. The meaning of the term 'speed' will, in each particular case, be clear from the context.

The use of the definition of velocity is illustrated for an important case in the following example.

Example 7 A particle is moving in a circle of radius c and centre O as shown in Fig. 2.6. At time t the particle is at P, where OP makes an angle $\theta = \theta(t)$ with a fixed line ON. Determine $\mathbf{v}$ and $|\mathbf{v}|$ in terms of $\dot{\theta}$.

Let Q be the position of the particle at time $t + \delta t$, so that $\overrightarrow{PQ} = \boldsymbol{\delta}\mathbf{r}$.
Let $\hat{\boldsymbol{\theta}}$ be a unit vector perpendicular to OP pointing in the direction in which

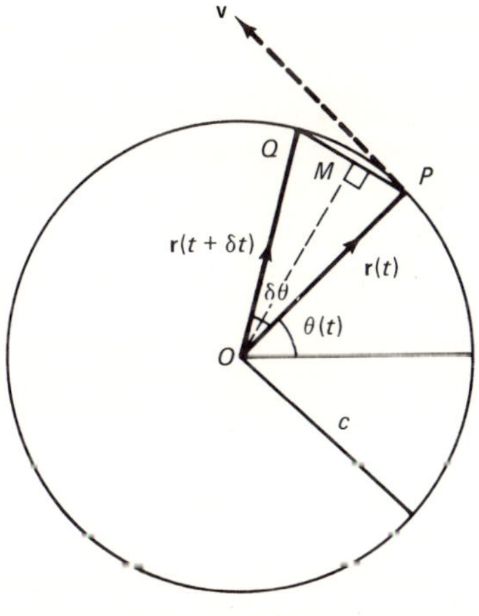

Fig. 2.6 Sketch illustrating Example 7.

θ increases. The magnitude of $\boldsymbol{\delta}\mathbf{r}$ is $PQ = 2PM = 2c\sin(\frac{1}{2}\delta\theta)$ where $\delta\theta = \theta(t+\delta t) - \theta(t)$. For small values of $\delta\theta$,

$$2c\sin(\tfrac{1}{2}\delta\theta) \approx 2c\cdot\frac{\delta\theta}{2} = c\delta\theta,$$

and the direction of $\boldsymbol{\delta}\mathbf{r}$ is close to that of $\hat{\boldsymbol{\theta}}$. Hence $\boldsymbol{\delta}\mathbf{r} \approx c\delta\theta\hat{\boldsymbol{\theta}}$.

$$(2.15) \quad \Rightarrow \quad \mathbf{v} = \lim_{\delta t\to 0}\left[\frac{c\delta\theta}{\delta t}\cdot\hat{\boldsymbol{\theta}}\right].$$

Hence

$$\dot{\mathbf{r}} = \mathbf{v} = c\dot{\theta}\hat{\boldsymbol{\theta}}. \tag{2.16}$$

Since $\hat{\boldsymbol{\theta}}$ is a unit vector, it follows that

$$|\mathbf{v}| = c\dot{\theta}. \tag{2.17}$$

There is a trap connected with velocity and speed into which beginners often fall. Let us adopt the standard convention of writing r for $|\mathbf{r}|$ and v for $|\mathbf{v}|$. Since $\mathbf{v} = \dot{\mathbf{r}}$ and $v = |\dot{\mathbf{r}}|$, it is often assumed that $v = |\dot{r}|$, *but this is wrong*. The error can be understood by considering Example 7 where r has the constant value c for all t, so that $\dot{r}$ (and so $|\dot{r}|$) is 0 for all t. However $v = |\dot{\mathbf{r}}|$ is obviously not zero; its value is given by (2.17). It takes a moment to see that the mathematical reason why $|\dot{\mathbf{r}}|$ and $|\dot{r}|$ are different is that the operations on a vector of forming its magnitude, and differentiating, do not commute. The existence of this trap is also a warning that it is very important to use a notation that distinguishes clearly between a vector such as $\mathbf{r}$ and a scalar such as r.

In Example 7, $\mathbf{v}$ is not a constant vector. Its direction is changing because the direction of $\hat{\boldsymbol{\theta}}$ is changing as P moves around the circle; its magnitude is also changing, unless $\dot{\theta}$ is constant. Thus, in a general case, we must expect $\mathbf{v}$ to depend on t, and we shall follow our usual practice of writing $\mathbf{v}(t)$ when we wish to emphasize this dependence.

Suppose that in a general case the particle P has velocities $\mathbf{v}(t)$ and $\mathbf{v}(t+\delta t)$ at times t and $t+\delta t$. By an obvious extension of the notation used in (2.14) we define $\boldsymbol{\delta}\mathbf{v}$ by

$$\boldsymbol{\delta}\mathbf{v} = \mathbf{v}(t+\delta t) - \mathbf{v}(t). \tag{2.18}$$

By a natural generalization of the process used to define velocity, we consider the limit of $\dfrac{\boldsymbol{\delta}\mathbf{v}}{\delta t}$ as δt approaches zero. The value of this limit is the *acceleration* of P, a vector denoted by $\mathbf{a}$, or by $\dot{\mathbf{v}}$, $\dfrac{\mathrm{d}\mathbf{v}}{\mathrm{d}t}$, $\ddot{\mathbf{r}}$ or $\dfrac{\mathrm{d}^2\mathbf{r}}{\mathrm{d}t^2}$. Thus

$$\mathbf{a} = \frac{\mathrm{d}\mathbf{v}}{\mathrm{d}t} = \dot{\mathbf{v}} = \frac{\mathrm{d}^2\mathbf{r}}{\mathrm{d}t^2} = \ddot{\mathbf{r}} = \lim_{\delta t\to 0}\left(\frac{\boldsymbol{\delta}\mathbf{v}}{\delta t}\right) = \lim_{\delta t\to 0}\left(\frac{\mathbf{v}(t+\delta t) - \mathbf{v}(t)}{\delta t}\right). \tag{2.19}$$

Note that the same word acceleration is used as for motion along a straight line. However the context will always indicate whether the motion is in three dimensions, in which case (2.19) defines the acceleration as a vector, or along a straight line, in which case (2.2) defines it as a scalar.

Example 8 Determine the acceleration of the particle in Example 7.

As in Example 7 let Q be the position of the particle at time $(t + \delta t)$. Also, as shown in Fig. 2.7, let $\hat{\mathbf{r}}(t)$ and $\hat{\boldsymbol{\theta}}(t)$ denote unit vectors along and perpendicular to OP, respectively, with $\hat{\mathbf{r}}(t + \delta t)$ and $\hat{\boldsymbol{\theta}}(t + \delta t)$ being their corresponding values at Q.

As already noted after Example 7, $\hat{\boldsymbol{\theta}}$ changes with t, and so does $\hat{\mathbf{r}}$. Consider the change $\boldsymbol{\delta}\hat{\mathbf{r}}$ in $\hat{\mathbf{r}}$ between t and $t + \delta t$, where $\boldsymbol{\delta}\hat{\mathbf{r}} = \hat{\mathbf{r}}(t + \delta t) - \hat{\mathbf{r}}(t)$. As Fig. 2.7 shows, $\boldsymbol{\delta}\hat{\mathbf{r}}$ is represented by $\overrightarrow{MN}$. Now, when δt is small, $\overrightarrow{MN}$ is nearly parallel to $\hat{\boldsymbol{\theta}}$, and, since $PM = PN = 1$, its length is $2\sin(\frac{1}{2}\delta\theta)$, which, when $\delta\theta$ is small, is approximately $2(\frac{1}{2}\delta\theta) = \delta\theta$. Thus

$$\boldsymbol{\delta}\hat{\mathbf{r}} \approx \delta\theta \, . \, \hat{\boldsymbol{\theta}}$$

$$\Rightarrow \quad \frac{\mathrm{d}}{\mathrm{d}t}\hat{\mathbf{r}} = \lim_{\delta t \to 0}\left(\frac{\boldsymbol{\delta}\hat{\mathbf{r}}}{\delta t}\right) = \dot{\theta}\hat{\boldsymbol{\theta}}. \tag{2.20}$$

By a similar argument (which the reader is advised to write out in full) using $\triangle PRS$ in Fig. 2.7, it can be shown that

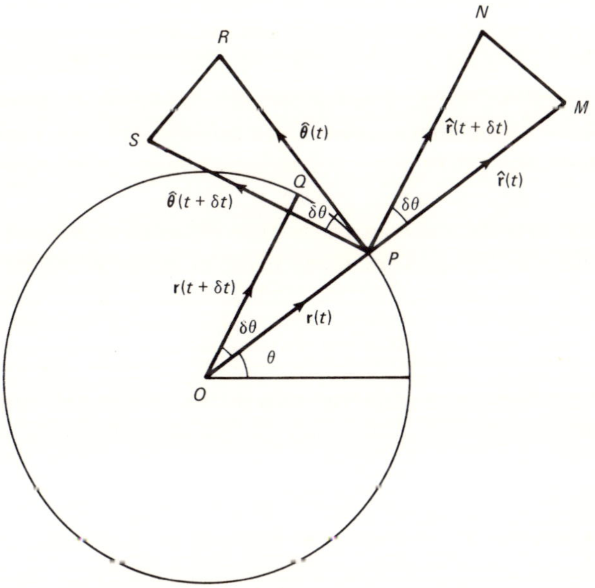

Fig. 2.7 Sketch illustrating Example 8.

$$\boldsymbol{\delta}\hat{\boldsymbol{\theta}} \approx -\delta\theta\,.\,\hat{\mathbf{r}}$$

$$\Rightarrow \quad \frac{\mathrm{d}}{\mathrm{d}t}\hat{\boldsymbol{\theta}} = \lim_{\delta t \to 0}\left(\frac{\boldsymbol{\delta}\hat{\boldsymbol{\theta}}}{\delta t}\right) = -\dot{\theta}\hat{\mathbf{r}}. \tag{2.21}$$

Let $\mathbf{r}$, represented by $\overrightarrow{OP}$, be the position vector of P. Since the length of OP is the constant value c, it is clear that $\mathbf{r} = c\hat{\mathbf{r}}$. Thus $\boldsymbol{\delta}\mathbf{r} = c\boldsymbol{\delta}\hat{\mathbf{r}} \approx c\delta\theta\hat{\boldsymbol{\theta}}$ by (2.20). Hence $\dot{\mathbf{r}} = c\dot{\theta}\hat{\boldsymbol{\theta}}$, as obtained by another method in Example 7.

We can obtain the acceleration $\mathbf{a} = \dfrac{\mathrm{d}}{\mathrm{d}t}\dot{\mathbf{r}}$ either from first principles, using methods similar to those used to derive (2.16) and (2.20), or we can simply differentiate (2.16). By the product rule

$$\mathbf{a} = \frac{\mathrm{d}}{\mathrm{d}t}(c\dot{\theta}\,.\,\hat{\boldsymbol{\theta}}) = c\left(\frac{\mathrm{d}}{\mathrm{d}t}\dot{\theta}\right)\hat{\boldsymbol{\theta}} + c\dot{\theta}\left(\frac{\mathrm{d}}{\mathrm{d}t}\hat{\boldsymbol{\theta}}\right) = c\ddot{\theta}\hat{\boldsymbol{\theta}} - c\dot{\theta}^2\hat{\mathbf{r}}, \tag{2.22}$$

where we have used (2.21) to determine $\dfrac{\mathrm{d}\hat{\boldsymbol{\theta}}}{\mathrm{d}t}$.

An interesting special case of the motion discussed in Examples 7 and 8 arises when $\dot{\theta} = \omega$, where ω is a constant. Then, using (2.16) and (2.22),

$$\mathbf{v} = c\omega\hat{\boldsymbol{\theta}}; \quad \mathbf{a} = -c\omega^2\hat{\mathbf{r}} = -\frac{v^2}{c}\hat{\mathbf{r}}, \tag{2.23}$$

since $\ddot{\theta} = 0$ when $\dot{\theta}$ is constant. The result $\mathbf{a} = -(v^2/c)\,\hat{\mathbf{r}}$ follows from $\mathbf{v} = c\omega\hat{\boldsymbol{\theta}}$ since this means $v = |\mathbf{v}| = c\omega$. Remember that $\hat{\boldsymbol{\theta}}$ is a unit vector so that $|\hat{\boldsymbol{\theta}}| = 1$. The acceleration is directed towards the centre of the circle along which P is moving because that is the direction of changes in the velocity. The constant ω is known as the *angular speed.* Its dimension is T^{-1} so that its units are s^{-1}. Since $\omega = \dot{\theta}$ and since θ is measured in radians, an angular speed of for example $\pi\ \mathrm{s}^{-1} \approx 3{\cdot}142\ \mathrm{s}^{-1}$ means that P takes 2 s to travel completely round the circle.

In the book *Vectors* in this series, it is shown how vectors can be expressed in terms of their *components* with respect to a *basis.* In Example 8 for example it was shown that $(-c\dot{\theta}^2, c\ddot{\theta})$ were the components of $\mathbf{a}$ with respect to the basis $(\hat{\mathbf{r}}, \hat{\boldsymbol{\theta}})$. There are only two vectors in this basis because the motion is in a plane.

More generally we want to be able to describe motion in three dimensions, so that a basis must contain three vectors. The most useful basis is illustrated in Fig. 2.8 and is known as the *cartesian basis*, usually denoted by $\{\mathbf{i}, \mathbf{j}, \mathbf{k}\}$, although the notation $(\mathbf{e}_1, \mathbf{e}_2, \mathbf{e}_3)$ with $\mathbf{e}_1 = \mathbf{i}$, etc., is also useful. Each of $\mathbf{i}$, $\mathbf{j}$, $\mathbf{k}$ is a unit vector, and each of them is perpendicular to the other two. Using the notation of the scalar (or dot) product, these properties can be expressed by the equations

$$\mathbf{i}\,.\,\mathbf{i} = \mathbf{j}\,.\,\mathbf{j} = \mathbf{k}\,.\,\mathbf{k} = 1; \quad \mathbf{j}\,.\,\mathbf{k} = \mathbf{k}\,.\,\mathbf{i} = \mathbf{i}\,.\,\mathbf{j} = 0. \tag{2.24}$$

Any vector $\mathbf{r}$ has components with respect to $\{\mathbf{i}, \mathbf{j}, \mathbf{k}\}$. If these are denoted by (x, y, z) we have

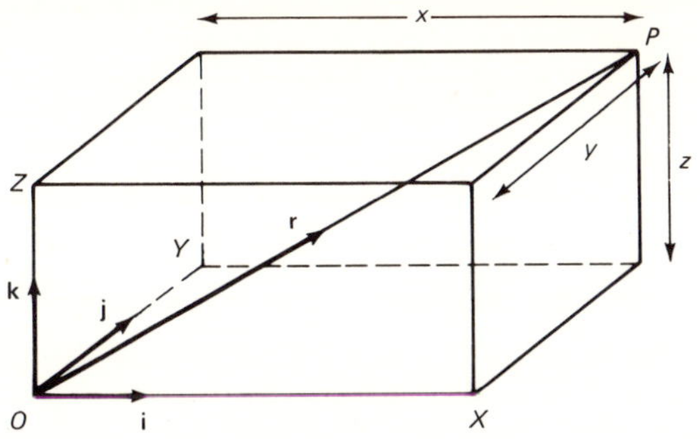

Fig. 2.8 Sketch defining the cartesian basis $\{\mathbf{i}, \mathbf{j}, \mathbf{k}\}$ and the components (x, y, z) of $\mathbf{r}$ with respect to this basis. Note that $x = OX$, $y = OY$, $z = OZ$.

$$r = x\mathbf{i} + y\mathbf{j} + z\mathbf{k}. \tag{2.25}$$

It follows from (2.24) and (2.25) that $\mathbf{r}\,.\,\mathbf{i} = x$, etc.; this is also illustrated in Fig. 2.8.

Now suppose that $\mathbf{r}$ is the position vector of a point P which is moving in space. Since $(\mathbf{i}, \mathbf{j}, \mathbf{k})$ are constant vectors, that is they have constant directions and lengths, unlike $\hat{\mathbf{r}}$, $\hat{\boldsymbol{\theta}}$, changes in $\mathbf{r}$ are represented entirely by changes in (x, y, z). Thus

$$\begin{aligned}\boldsymbol{\delta}\mathbf{r} &= \mathbf{r}(t + \delta t) - \mathbf{r}(t)\\ &= [x(t + \delta t) - x(t)]\mathbf{i} + [y(t + \delta t) - y(t)]\mathbf{j} + [z(t + \delta t) - z(t)]\mathbf{k}\\ &= \delta x\mathbf{i} + \delta y\mathbf{j} + \delta z\mathbf{k}.\end{aligned}$$

Hence, from (2.15),

$$\mathbf{v} = \dot{\mathbf{r}} = \dot{x}\mathbf{i} + \dot{y}\mathbf{j} + \dot{z}\mathbf{k}, \tag{2.26}$$

so that $(\dot{x}, \dot{y}, \dot{z})$ are the components of the velocity $\mathbf{v}$ with respect to the basis $(\mathbf{i}, \mathbf{j}, \mathbf{k})$. Note that $(\dot{x}, \dot{y}, \dot{z})$ are scalars which can be positive or negative. Similarly the acceleration $\mathbf{a}$ is given by

$$\mathbf{a} = \dot{\mathbf{v}} = \ddot{\mathbf{r}} = \ddot{x}\mathbf{i} + \ddot{y}\mathbf{j} + \ddot{z}\mathbf{k}. \tag{2.27}$$

Example 9 A particle is moving so that its position vector $\mathbf{r}$ at time t satisfies $\mathbf{r} = (c\cos\omega t)\mathbf{i} + (c\sin\omega t)\mathbf{j}$, where c and ω are positive constants. Find its velocity $\mathbf{v}$ and acceleration $\mathbf{a}$.

Since

$$\frac{\mathrm{d}}{\mathrm{d}t}(c\cos\omega t) = c\frac{\mathrm{d}}{\mathrm{d}t}(\cos\omega t) = -c\omega\sin\omega t,$$

and

$$\frac{d}{dt}(c \sin \omega t) = c\frac{d}{dt}(\sin \omega t) = c\omega \cos \omega t,$$

it follows from (2.26) and (2.27) that

$$\begin{aligned} \mathbf{v} &= (-c\omega \sin \omega t)\mathbf{i} + (c\omega \cos \omega t)\mathbf{j}; \\ \mathbf{a} &= (-c\omega^2 \cos \omega t)\mathbf{i} + (-c\omega^2 \sin \omega t)\mathbf{j}. \end{aligned} \tag{2.28}$$

The motion in Example 9 takes place in the plane $z = 0$, with $x = c \cos \omega t$, $y = c \sin \omega t$. Hence

$$x^2 + y^2 = c^2(\cos^2 \omega t + \sin^2 \omega t) = c^2,$$

so that P moves round a circle centre O radius c, just as in the motion considered in Examples 7 and 8, pp. 17 and 19. In fact the motion is such that the angle θ between OP and $\mathbf{i}$ is ωt (please verify); hence $\dot{\theta} = \omega$. Thus the motion in Example 9 is the special case of that in Examples 7 and 8 for which (2.23) gives the velocity and acceleration. In other words (2.23) and (2.28) should be equivalent. It is left as an exercise for the reader to show that this is so. (It is suggested that you express $\hat{\mathbf{r}}$ and $\hat{\boldsymbol{\theta}}$ in terms of $\mathbf{i}$ and $\mathbf{j}$.)

In this book we adopt the practice that the dimension of a vector is the dimension of its magnitude. For example, since the dimension of speed is LT^{-1}, so also is it the dimension of velocity $\mathbf{v}$. Consider (2.26), where the left-hand side has dimension LT^{-1}. For dimensional consistency the right-hand side of (2.26) must also have dimension LT^{-1} (see §1.3). Now $\dot{x} = \dfrac{dx}{dt}$ has this dimension but $\mathbf{i}$ has no dimension, since $|\mathbf{i}| = 1$, not 1 m—hence $\dot{x}\mathbf{i}$ has the required dimension; so also do $\dot{y}\mathbf{j}$ and $\dot{z}\mathbf{k}$. Since c and ω in Example 9 have dimensions L and T^{-1} respectively, it can also be seen that (2.28) is dimensionally consistent. Finally, note that just as we can consider a speed of 5 ms^{-1}, so also can we consider a velocity of $(3\mathbf{i} + 4\mathbf{j} - 2\mathbf{k})$ ms^{-1}.

The next example shows that some of the techniques developed in §2.2 for motion in a straight line can also be useful when dealing with more complicated motions.

Example 10 A particle is moving with acceleration $-g\mathbf{j}$, where $\mathbf{j}$ is vertically upwards and g is a constant. At $t = 0$, $\mathbf{r} = \mathbf{0}$ and $\mathbf{v} = V \cos \alpha \mathbf{i} + V \sin \alpha \mathbf{j}$, where $\mathbf{i}$ is horizontal and V and α are constants. Determine the velocity and position of the particle for all t.

$$(2.27) \quad \Rightarrow \quad \ddot{x}\mathbf{i} + \ddot{y}\mathbf{j} + \ddot{z}\mathbf{k} = -g\mathbf{j} \quad \Rightarrow \quad \ddot{x} = 0, \ddot{y} = -g, \ddot{z} = 0.$$

Hence, as in Example 2, p. 10,

$$\dot{x} = \text{constant}, \ \dot{y} = -gt + \text{constant}, \ \dot{z} = \text{constant}.$$

We are given that, at $t = 0$,

$$\ddot{x} = V\cos\alpha, \quad \dot{y} = V\sin\alpha, \quad \dot{z} = 0$$

$$\Rightarrow \quad \dot{x} = V\cos\alpha, \quad \dot{y} = V\sin\alpha - gt, \quad \dot{z} = 0.$$

Similarly we integrate once more and use $x = y = z = 0$ at $t = 0$ to obtain

$$x = Vt\cos\alpha, \quad y = Vt\sin\alpha - \tfrac{1}{2}gt^2, \quad z = 0.$$

These results may be written in vector notation as

$$\mathbf{v} = V\cos\alpha\mathbf{i} + (V\sin\alpha - gt)\mathbf{j}; \quad \mathbf{r} = Vt\cos\alpha\mathbf{i} + (Vt\sin\alpha - \tfrac{1}{2}gt^2)\mathbf{j}. \qquad (2.29)$$

2.4 Relative motion

We now have the machinery to discuss an important class of problems, namely those in which the interest is in the motion of one moving body relative to another. A typical situation is that of two aircraft moving in different directions at different speeds. Here our main interest is whether or not collision occurs. The following discussion illustrates one of several methods available for solving this type of problem.

Suppose the two bodies are labelled P and Q and that they have position vectors $\mathbf{p}$ and $\mathbf{q}$ which both change with t since P and Q are both moving. As illustrated in Fig. 2.9 the position vector of P relative to Q is $\mathbf{s} = \mathbf{p} - \mathbf{q}$. In this class of problems we are interested in how $\mathbf{s}$ varies with t. Note that the rate of change of $\mathbf{s}$ is $\dot{\mathbf{s}} = \dot{\mathbf{p}} - \dot{\mathbf{q}}$, which is the difference between the velocities of P

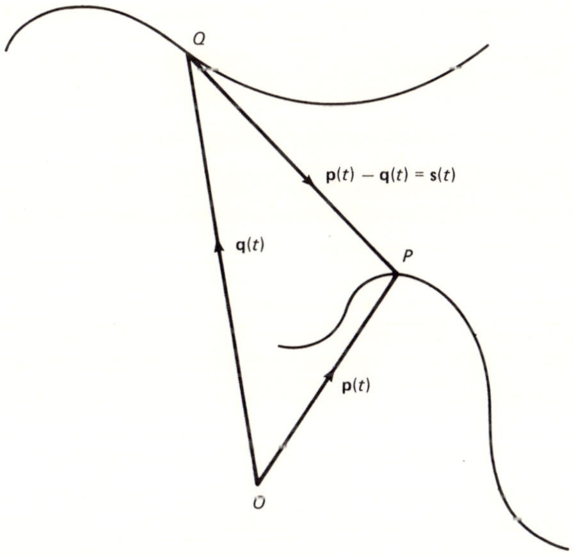

Fig. 2.9 Sketch illustrating the discussion on relative motion.

and Q and is known as the *relative velocity* or, more precisely, the velocity of P relative to Q. In effect we have changed our origin from the fixed point O to the moving point Q, and $\mathbf{s}$ and $\dot{\mathbf{s}}$ are simply the position vector and velocity of P with respect to the new moving origin. Whatever our viewpoint, it is clear that $\mathbf{s}(t)$ gives complete information about the relative motion of P and Q.

The use of these ideas is illustrated in the next example.

Example 11 An aircraft has constant speed 90 ms^{-1} and is flying in a straight line parallel to $(-\mathbf{i}+2\mathbf{j}+2\mathbf{k})$. A helicopter has constant speed 15 ms^{-1} and is flying in a straight line parallel to $\mathbf{j}$. Initially the aircraft is at the origin and the helicopter is 1000 m from the origin along a line parallel to $(-2\mathbf{i}+3\mathbf{j}+4\mathbf{k})$. Show that collision is likely, and find when it will occur if no avoiding action is taken.

Denote the aircraft by P and let its position vector relative to the origin after T s be $\mathbf{p}$. Define Q and $\mathbf{q}$ similarly for the helicopter.

$$|-\mathbf{i}+2\mathbf{j}+2\mathbf{k}| = 3 \quad\Rightarrow\quad \dot{\mathbf{p}} = 30(-\mathbf{i}+2\mathbf{j}+2\mathbf{k})\ \text{ms}^{-1}$$

$$\Rightarrow\quad \mathbf{p} = 30T(-\mathbf{i}+2\mathbf{j}+2\mathbf{k})\ \text{m}.$$

$$\dot{\mathbf{q}} = 15\mathbf{j}\ \text{ms}^{-1}.$$

$$|-2\mathbf{i}+3\mathbf{j}+4\mathbf{k}| = \sqrt{29} \quad\Rightarrow\quad \mathbf{q} = \left[15T\mathbf{j}+\frac{1000}{\sqrt{29}}(-2\mathbf{i}+3\mathbf{j}+4\mathbf{k})\right]\ \text{m}$$

$$\Rightarrow\quad (\mathbf{p}-\mathbf{q}) = \left(15T-\frac{1000}{\sqrt{29}}\right)(-2\mathbf{i}+3\mathbf{j}+4\mathbf{k})\ \text{m}.$$

Since $(\mathbf{p}-\mathbf{q})$ is the position vector of P relative to Q, collision is likely if there is a value of T for which $\mathbf{p}-\mathbf{q}=\mathbf{0}$. It can be seen that this is so when $15T = (1000/\sqrt{29})$; that is, after about 12·4 s.

The methods used to analyse the relative motion of two moving bodies can also be used to determine the motion of one body moving in a medium which is itself moving, such as a boat sailing on a moving sea or, as in the following example, a helicopter flying in a wind.

Example 12 A ship is sailing due north at a constant speed of 20 km h^{-1}. At noon the ship is sighted 5 km due east of a shore station at which there is a helicopter. There is wind of speed 50 km h^{-1} blowing from the direction α east of south, where α is acute and $\tan\alpha = 3/4$. The helicopter takes off T_0 hours after noon to intercept the ship and flies at a constant speed of 60 km h^{-1} relative to the wind in a constant direction β east of south relative to the wind, where β is chosen to ensure interception. Find

(i) the flight time when $T_0 = 0$;

(ii) the least possible flight time and the corresponding value of T_0.

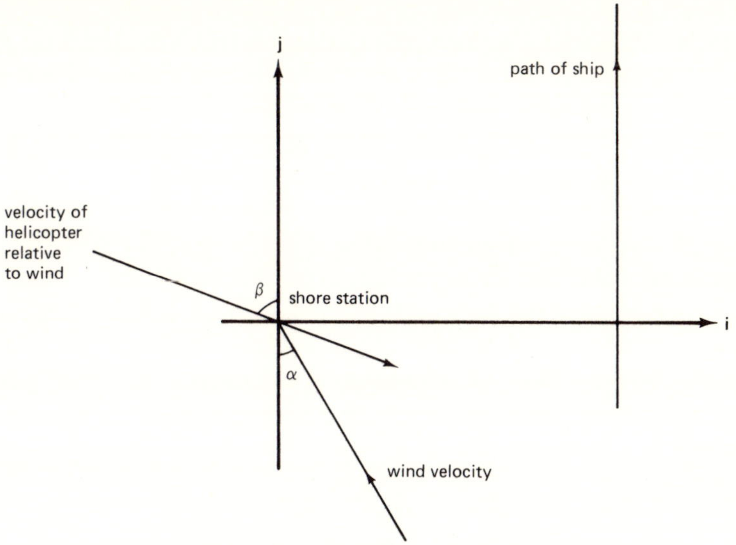

Fig. 2.10 Sketch illustrating Example 12.

The situation is illustrated in Fig. 2.10. Unit vectors **i** and **j** are taken east and north respectively.

Let **p** and **q** denote the position vectors of the ship and helicopter respectively relative to the shore station at T hours after noon. We are given that

$$\mathbf{p} = (5\mathbf{i} + 20T\mathbf{j})\ \text{km}.$$

Let **v** be the velocity of the wind.

$$\tan\alpha = 3/4,\ \alpha \text{ acute} \quad \Rightarrow \quad \sin\alpha = 3/5,\ \cos\alpha = 4/5.$$

$$\Rightarrow \quad \mathbf{v} = 50(-\sin\alpha\mathbf{i} + \cos\alpha\mathbf{j})\ \text{km h}^{-1} = (-30\mathbf{i} + 40\mathbf{j})\ \text{km h}^{-1}.$$

We are given that $(\dot{\mathbf{q}} - \mathbf{v})$, the velocity of the helicopter relative to the wind, has magnitude 60 km h^{-1} and direction β east of south.

$$\Rightarrow \quad (\dot{\mathbf{q}} - \mathbf{v}) = 60(\sin\beta\mathbf{i} - \cos\beta\mathbf{j})\ \text{km h}^{-1}$$

$$\Rightarrow \quad \dot{\mathbf{q}} = [(60\sin\beta - 30)\mathbf{i} + (-60\cos\beta + 40)\mathbf{j}]\ \text{km h}^{-1}.$$

At noon $\mathbf{q} = 0$. Hence T hours after noon, where $T > T_0$, the helicopter has been flying for $(T - T_0)$ hours

$$\Rightarrow \quad \mathbf{q} = [(60\sin\beta - 30)\mathbf{i} + (-60\cos\beta + 40)\mathbf{j}](T - T_0)\ \text{km}.$$

Interception occurs when $\mathbf{p} = \mathbf{q}$.

$$\Rightarrow \quad 5 = (T - T_0)(60\sin\beta - 30); \quad 20T = (T - T_0)(-60\cos\beta + 40).$$

(i) $T_0 = 0 \quad \Rightarrow \quad 5 = 30T(2\sin\beta - 1);\ 1 = 2 - 3\cos\beta$

$$\Rightarrow \quad \cos\beta = 1/3 \quad \Rightarrow \quad \sin\beta = (2\sqrt{2/3})$$

$$\Rightarrow \quad T = (8\sqrt{2} - 6)^{-1} \approx 0{\cdot}188$$

$$\Rightarrow \quad \text{flight time} \approx 0{\cdot}188 \text{ hours} \approx 11{\cdot}3 \text{ minutes.}$$

(ii) Flight time is $(T - T_0)$ hours $= [6(2\sin\beta - 1)]^{-1}$ hours. This is least when $6(2\sin\beta - 1)$ is greatest; that is when $\sin\beta = 1$.

$$\sin\beta = 1 \quad \Rightarrow \quad (T - T_0) \text{ hours} = (1/6) \text{ hours} = 10 \text{ minutes.}$$

Also $(T - T_0) = 1/6 \quad \Rightarrow \quad T = T_0 + 1/6$

$$\Rightarrow \quad 20(T_0 + 1/6) = (1/6)(40), \text{ since } \sin\beta = 1 \quad \Rightarrow \quad \cos\beta = 0$$

$$\Rightarrow \quad T_0 = 1/6 \text{ hours} = 10 \text{ minutes.}$$

Exercise 2

1 A particle moves in a straight line Ox in the direction of increasing x so that its distance x from O and its speed v at any instant satisfy

$$e^{x/a} = 1 + v/V,$$

where a and V are positive constants. Show that the acceleration of the particle is $v(v + V)/a$.

2 A particle starts from O with speed U and travels in a straight line. After a time t its acceleration is $a(1 + kt)^{-2}$, where a and k are positive constants. What are the dimensions of a and k? Find the speed of the particle at time t, and show that this can never exceed $(U + a/k)$. Show that the distance of the particle from O at time t is

$$\left(U + \frac{a}{k}\right)t - \frac{a}{k^2}\ln(1 + kt).$$

3 Two trains P and Q travel by the same route from rest at station A to rest at station B. Train P has constant acceleration f for the first third of the time, constant speed for the second third, and constant acceleration $-f$ for the final third. Train Q has constant acceleration f for the first third of the distance, constant speed for the second third, and constant acceleration $-f$ for the final third. Show that the ratio of the times taken by the two trains is $\sqrt{(27/25)} \approx 1{\cdot}04$.

4 A particle moves along a straight line so that at time t its distance x from a fixed point O satisfies $x = c\cos(\omega t + \alpha)$, where c, ω and α are positive constants, with $\alpha < 3$. Its maximum speed is 2 ms^{-1} and its maximum distance from O is 0·5 m. When $t = 1$ s, the particle is 0·25 m from O and travelling towards O. Find ω, c and α, giving the value of α in radians to two decimal places.

5 An artificial satellite is fired vertically from the Earth's surface. When its distance from the centre of the Earth is x its speed $\dot{x}$ is given by

$$\dot{x} = W(1 + Xx^{-1})^{1/2},$$

where X and W are positive constants. Show that this equation is satisfied when x and t are related by

$$W(t - t_0)/X = \left(\frac{x}{X}\right)^{1/2}\left(1 + \frac{x}{X}\right)^{1/2} - \ln\left[\left(\frac{x}{X}\right)^{1/2} + \left(1 + \frac{x}{X}\right)^{1/2}\right],$$

where t_0 is a constant. Plot an accurate graph of (Wt/X) against (x/X) for the case when $t_0 = 0$, and hence estimate the value of (x/X) when $(Wt/X) = 1$.

6 A particle is moving so that at time t its speed is v. Observations give the data below.

t(s)	0	0·05	0·10	0·15	0·20	0·25	0·30	0·35	0·40
$v(\mathrm{ms}^{-1})$	0	0·35	0·59	0·68	0·74	0·79	0·79	0·81	0·82

Plot these data on a graph and draw the curve which appears to be the best fit. By counting squares, estimate the distance travelled between $t = 0$ s and $t = 0{\cdot}40$ s. Also estimate this distance by using the trapezium rule.

Given that theory suggests that v and t are related by

$$v = (g/k)(1 - e^{-kt}),$$

where g and k are positive constants, use your graph to obtain approximate values of g and k. Hence, by integration, obtain a third estimate of the distance travelled.

7 A point is travelling in the positive direction on the x-axis with acceleration proportional to the square of its speed v. At time $t = 0$ it passes through the origin with speed gT and with acceleration g. Show that $v' = v/(gT^2)$, and hence obtain an expression for v in terms of x, g and T.

Prove that at time t

$$x = gT^2 \ln [T/(T - t)].$$

8 A particle P moves in a plane in such a way that its distance from a fixed point O has the constant value a. The line OP rotates in the counter-clockwise sense with (variable) angular speed ω. By expressing the position vector $\mathbf{r}$ of P relative to O in the form

$$\mathbf{r} = a(\mathbf{i}\cos\theta + \mathbf{j}\sin\theta),$$

where $\mathbf{i}$ and $\mathbf{j}$ are perpendicular unit vectors, prove that the velocity $\dot{\mathbf{r}}$ of P is perpendicular to OP and that

$$\ddot{\mathbf{r}} = (\dot{\omega}/\omega)\dot{\mathbf{r}} - \omega^2\mathbf{r}.$$

Deduce that the magnitude of the acceleration is $a\sqrt{(\dot{\omega}^2 + \omega^4)}$.

9 A particle moving in a plane has acceleration at time t equal to

$$-\omega^2(a\mathbf{i}\cos\omega t + b\mathbf{j}\sin\omega t),$$

where a, b and ω are positive constants. At time $t = 0$ the particle has position vector $a\mathbf{i}$ and velocity $b\omega\mathbf{j}$. Find the position vector of the particle at time t, and deduce that its path is an ellipse.

10 At time t the position vector $\mathbf{r}$ of a particle relative to a fixed origin O satisfies $\mathbf{r} = c(\mathbf{i}\sin\omega t + \mathbf{j}\cos\omega t) + Ut\mathbf{k}$, where U, c and ω are constants. Describe the motion of the particle, and determine its velocity and acceleration for all t.

11 A gramophone turntable is rotating with constant angular speed ω about an axis through its centre O. A fly walks outwards along a radius through O with constant speed V relative to the turntable. Find the magnitude of the fly's acceleration when it is at a distance r from O.

12 Particles A and B start at time $t = 0$ from points with position vectors $(5\mathbf{i} + 13\mathbf{j})$ m and $(7\mathbf{i} + 5\mathbf{j})$ m respectively. The velocities of A and B are constant and equal to $(3\mathbf{i} - 5\mathbf{j})\ \mathrm{ms}^{-1}$ and $(2\mathbf{i} - \mathbf{j})\ \mathrm{ms}^{-1}$ respectively. Determine the velocity of A relative to B, and deduce that the particles collide. Find the position vector of the point of collision.

13 At time $t = 0$ a particle A is at the origin and a particle B is at the point with position vector $(5\mathbf{i} - 10\mathbf{j} - 12\mathbf{k})$ m. The particles A and B have the constant velocities $2\mathbf{i}\ \mathrm{ms}^{-1}$ and $(4\mathbf{i} + 4\mathbf{j} + 5\mathbf{k})\ \mathrm{ms}^{-1}$ respectively. Show that the least distance between A and B in the motion is about 9·43 m.

14 Two straight roads OE and ON run east–west and north–south respectively. Two motor cars are travelling along these roads with uniform speeds. One car X is travelling along OE towards O with speed 42 km h^{-1} and the other car Y is travelling along ON towards O with speed 56 km h^{-1}. At $t = 0$ the distances of X and Y from O are 0·5 km and 0·75 km respectively. Find the closest distance between X and Y.

15 An aeroplane, whose speed in still air is 800 km h^{-1}, flies at a constant height and describes a horizontal circuit in the shape of an equilateral triangle of side 200 km. The vertices of the triangle are A, B, C, where A is north of BC, and B is due west of C. A wind of speed 100 km h^{-1} is blowing from the west. Show that a complete circuit travelled in the anti-clockwise sense takes about 0·76 h.

16 An aeroplane has constant speed 135 ms^{-1} in the direction of $(7\mathbf{i} - 4\mathbf{j} - 4\mathbf{k})$ and is initially at the point with position vector $300(\mathbf{i} + 4\mathbf{j} + 4\mathbf{k})$ m. A second aeroplane, initially at the origin, has constant speed u parallel to $\mathbf{i}$.

(i) Show that the aeroplanes are on a collision course if u has a certain value, and determine when collision is likely in this case.

(ii) Show that when $u = 135 \text{ ms}^{-1}$ the minimum distance between the aeroplanes is $200\sqrt{2}$ m.

17 A bus starts from rest and moves along a straight road with constant acceleration f until its speed is V; it then continues at constant speed V. When the bus starts, a car is at a distance b behind the bus and is moving in the same direction with constant speed U. Find the distance of the car behind the bus at time t after the bus has started for

$$\text{(i) } 0 \leqslant t < V/f, \qquad \text{and (ii) } t \geqslant V/f.$$

Show that the car cannot overtake the bus during the period $0 \leqslant t < V/f$ unless $U^2 > 2bf$.

Find the least distance between the car and the bus in the case when $U^2 < 2bf$ and $U < V$. State briefly what will happen if $U^2 < 2bf$ and $U > V$.

18 The ends A and B of a beam of length $2a$ are in contact with a vertical wall and a horizontal floor respectively. The vertical plane through the beam is perpendicular to the plane of the wall. If A is moved up the wall with constant speed V, find the speed of B when AB makes an angle $(\pi/6)$ with the horizontal. Find also the acceleration of B at this instant.

19 At time $t = 0$ a boat S sails due east with constant speed U from a point A. Simultaneously a second boat T sails with constant speed λU $(\lambda < 1)$ from a point B which is at a distance d due south of A. Given that the velocities of both boats remain constant, find the direction in which the boat T must sail in order to pass as close as possible to the boat S. Find the shortest distance between the boats and the value of t at the instant of closest approach.

20 A helicopter has to fly in a straight line from A to B and back, where B is a distance d due north of A. In still air the maximum speed of the helicopter is v. During the flight a wind with speed u $(u < v)$ is blowing from the south–east. In order that the helicopter moves along AB it is steered at an angle θ to $\overrightarrow{AB}$. Assuming that the helicopter maintains its maximum speed throughout the flight, show that $v\sqrt{2} = u \operatorname{cosec} \theta$, and that the speed of the helicopter (flying from A to B) relative to the Earth is $v \cos\theta + \frac{1}{2}u\sqrt{2}$.

Show that the time taken for the complete journey is

$$\frac{2d\sqrt{(v^2 - \frac{1}{2}u^2)}}{(v^2 - u^2)}.$$

3 Newton's laws of motion

3.1 Introduction

The purpose of this chapter is to introduce and explain the laws that enable us to obtain the acceleration of a particle. Once this has been done we can use the methods discussed in Chapter 2 to find the particle's velocity and position. The laws were formulated by Sir Isaac Newton and were published in 1687.

3.2 Force

The basic idea underlying Newton's laws is that a particle accelerates when, and only when, it is acted on by a *force*. We shall soon see that the word 'force' has a precise meaning in mechanics, unlike its use in everyday language, and that it is a vector quantity. Before giving a precise definition, it is useful to describe some forces that are important in mechanics.

When a ball is thrown into the air it falls back to the ground because the Earth exerts a force on it. This force is due to *gravity*, a phenomenon whose influence is universal in the sense that every particle exerts a gravitational force on every other particle. Gravity is also the origin of the force which enables the Earth to move round the Sun, and of the force which causes a body to have weight. In fact the weight of a body is defined to be the gravitational force exerted on it by the Earth. When we lift a body, our muscles have to exert a force sufficient to overcome its weight.

Another common and important force is *friction* between two surfaces in contact, which are moving, or trying to move, relative to one another. We are able to walk only because the ground exerts a frictional force on our feet in the direction of movement. Similarly, the acceleration of a car is caused by the frictional force exerted by the road on its tyres. Although we shall study friction in more detail later, we know from everyday experience that the magnitude of the frictional force between two touching surfaces depends on the nature of these surfaces. It is much easier to walk on a road than on ice because the road can exert a greater frictional force on our feet than can ice. For the same reason, much attention is devoted to the design of car tyres because of the large frictional force required between them and the road surface to avoid skidding.

Among other types of forces that we shall meet are *elastic* forces, such as those enabling a catapult to operate, and *air resistance*, so important in the design of aircraft and other bodies moving at high speed.

The magnitude of a force and the direction in which it acts both depend on its type. Later we shall learn how to give a mathematical description of these different types of forces.

For the moment, the basic point is that the motion of a body is determined by the net effect of all the forces acting on it.

3.3 Newton's laws of motion

We shall state Newton's laws of motion in modern language rather than use his words. There are three laws. Newton's First Law is:

When there are no forces acting on a particle, it moves with constant velocity. (3.1)

Newton's First Law is often stated as: 'When there are no forces acting on a particle, it moves with constant speed along a straight line'. By the work in §2.3, this is the same as (3.1), since a velocity is constant only when its magnitude and direction are separately constant. The First Law does not determine the value of the constant velocity with which a particle moves when no forces act on it. Indeed any constant velocity is consistent with the law. In particular, the constant velocity may be zero, in which case the particle is at rest.

It follows from Newton's First Law that when the acceleration of a particle is not zero, the total force acting on it cannot be zero. This fact is the basis of Newton's Second Law, which is:

The force on a particle is equal to the product of its mass and acceleration.

Note first that, as seen in §2.3, the acceleration of a particle is a vector. Newton's Second Law then shows that the force on a particle is a vector. Denoting this vector by **F**, Newton's Second Law can then be written

$$\mathbf{F} = m\mathbf{a}, \tag{3.2}$$

where m is the mass of the particle and **a** is its acceleration, defined in (2.19) as $\dot{\mathbf{v}}$, where the dot over **v** denotes its rate of change with respect to time.

In (3.2) the force **F** is the net force on the particle, so that if two (or more) forces act on the particle, then **F** is the vector sum of these two (or more) separate forces.

Equation (3.2) shows that the dimension of force is MLT^{-2} and that its units are $\mathrm{kg\,m\,s^{-2}}$. In fact force is such an important quantity that its unit, namely $1\ \mathrm{kg\,m\,s^{-2}}$, is given the special name of 1 newton (abbreviated to 1 N).

Example 1 A particle of mass m is moving so that at time t its position vector with respect to a fixed origin is $\mathbf{r} = \mathbf{V}t + \frac{1}{2}\mathbf{g}t^2$, where **V** and **g** are constant vectors. Find the force **F** on the particle. Find also the magnitude of **F** when $m = 0{\cdot}1$ kg and **g** has magnitude $9{\cdot}81\ \mathrm{ms^{-2}}$.

$$(2.15) \Rightarrow \mathbf{v} = \dot{\mathbf{r}} = \mathbf{V} + \mathbf{g}t.$$

$$(2.19) \Rightarrow \mathbf{a} = \dot{\mathbf{v}} = \mathbf{g}.$$

$$(3.1) \Rightarrow \mathbf{F} = m\mathbf{g}.$$

Hence $$|\mathbf{F}| = m|\mathbf{g}| = 0{\cdot}1 \times 9{\cdot}81\ \mathrm{N} = 0{\cdot}981\ \mathrm{N}.$$

Example 2 To a good approximation the Earth is moving round the Sun with constant speed in a circular path of radius $1{\cdot}5 \times 10^{11}$ m. Given that the mass of the Earth is approximately 6×10^{24} kg, find the magnitude and direction of the force exerted on the Earth.

The Earth takes ≈ 365 days $\approx 3{\cdot}15 \times 10^7$ s to move once round the Sun. Hence its speed

$$v \approx \frac{2\pi \times 1{\cdot}5 \times 10^{11}}{3{\cdot}15 \times 10^7}\ \mathrm{ms}^{-1} \approx 2{\cdot}99 \times 10^4\ \mathrm{ms}^{-1}.$$

By (2.23) its acceleration **a** is directed towards the Sun and has magnitude

$$\frac{(2{\cdot}99 \times 10^4)^2}{1{\cdot}5 \times 10^{11}}\ \mathrm{ms}^{-2} \approx 5{\cdot}96 \times 10^{-3}\ \mathrm{ms}^{-2}.$$

By (3.2) the force on the Earth has magnitude

$$[(6 \times 10^{24}) \times (5{\cdot}96 \times 10^{-3})]\ \mathrm{N} \approx 3{\cdot}6 \times 10^{22}\ \mathrm{N},$$

and is directed towards the Sun.

The third and last of Newton's laws concerns the forces which two particles exert on one another. Let the two particles be denoted by P and Q. Then Newton's Third Law is:

If P exerts a force **F** *on Q, then Q exerts a force* $-$**F** *on P.* (3.3)

In many books this law is often stated as: 'Action and reaction are equal and opposite'. We can think of the force **F** exerted on Q by P as the 'action', and the force $-$**F** exerted by Q on P as the 'reaction'. However this terminology suggests that the reaction is a consequence of the action, and this is not generally true.

Example 3 A horse is pulling a cart along a level road by means of light traces. Show on a clear sketch the forces causing the motion of (i) the horse, and (ii) the cart.

Criticize the statement: 'Although the horse is pulling the cart forwards, the cart is pulling the horse backwards with an equal and opposite force. Therefore there would be no motion except for the fact that the horse pulls first'.

The situation is sketched in Fig. 3.1. Let us suppose that the cart has two wheels in contact with the road along a line passing through A. To avoid unnecessary complications, the sketch shows only two legs of the horse, and the feet on these legs are in contact with the road at B and C. The traces are fastened to the cart at D and to the horse at E.

The forces acting on the horse are:

(a) its weight **W** acting vertically downwards;

(b) the force **R** exerted by the ground on the horse;

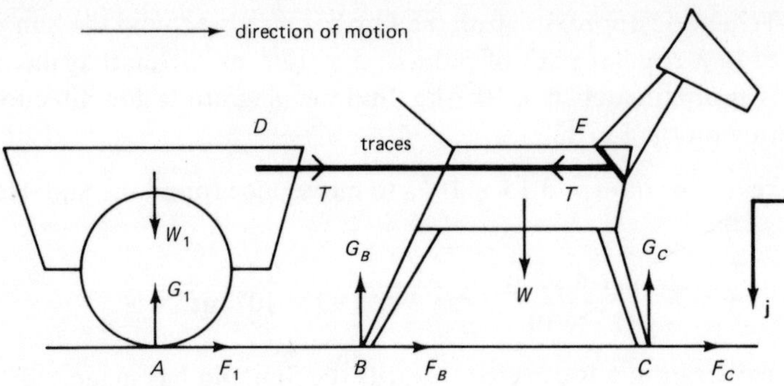

Fig. 3.1 Sketch illustrating Example 3.

(c) the force **T** exerted by the traces on the horse.

Let **i** and **j** be unit vectors in the directions shown on Fig. 3.1; thus **i** is in the direction of motion of the horse and cart, and **j** is vertically downwards. We express the forces (a), (b) and (c) in terms of their components with respect to **i** and **j**. For (a) and (c) this is easy. Thus $\mathbf{W} = W\mathbf{j}$, where W is the magnitude of the horse's weight, and $\mathbf{T} = -T\mathbf{i}$, where T is the magnitude of the force exerted by the traces and the minus sign is put in because **T** acts in the opposite direction to **i**. The direction of **R** is not parallel to **i** or **j**, that is **R** has non-zero components in both the horizontal and vertical directions. We therefore write $\mathbf{R} = F\mathbf{i} - G\mathbf{j}$. The *total* force on the horse is

$$\mathbf{W} + \mathbf{R} + \mathbf{T} = (F - T)\mathbf{i} + (W - G)\mathbf{j}.$$

The acceleration of the horse is horizontal and so is equal to $a\mathbf{i}$. Let m be the mass of the horse. Then from (3.2),

$$(F - T)\mathbf{i} + (W - G)\mathbf{j} = ma\mathbf{i}$$

$$\Rightarrow \quad F - T = ma, \quad W - G = 0. \qquad (3.4)$$

Thus, $G = W$ shows that the vertical component of **R**, the force exerted by the ground on the horse, balances the weight of the horse. The equation $F - T = ma$ shows that the value of a is determined by the values of F and T, and the horse can move its legs to control the value of F. If it wishes to increase its speed it moves its legs so that $F > T$. The value of F (and of $G = W$) is made up of contributions from each foot in contact with the ground.

Now consider the motion of the cart. The traces exert a pull on it. Since the traces are light their mass can be neglected. Hence, by applying (3.2) to the traces, the total force exerted on the traces is zero. Therefore the force exerted on the cart by the traces is $T\mathbf{i}$. The other forces exerted on the cart are its weight $W_1\mathbf{j}$ and the force $\mathbf{R}_1 = F_1\mathbf{i} - G_1\mathbf{j}$ exerted by the road on the cart. The acceleration of the cart is the same as that of the horse, namely $a\mathbf{i}$, so that, denoting the mass of the cart by m_1, equation (3.2) gives

$$(F_1 + T)\mathbf{i} + (W_1 - G_1)\mathbf{j} = m_1 a\mathbf{i}$$

$$\Rightarrow \quad F_1 + T = m_1 a, \quad W_1 - G_1 = 0. \tag{3.5}$$

Except for the fact that the cart, being inanimate, is unable to control the value of F_1, the interpretation of (3.5) is similar to that of (3.4). (But note that F_1 will normally be negative since it must act to turn the wheels in a clockwise sense.)

It is important to note that according to (3.3) there are forces $-\mathbf{R}$ and $-\mathbf{R}_1$ exerted *on* the ground by the horse and by the cart respectively. These forces have not been considered because they influence the motion of the ground, which is not relevant to the question.

Also note that by adding (3.4) and (3.5) we obtain

$$(F + F_1) = (m + m_1)a,$$

which is an equation for the motion of the horse and cart together.

The discussion shows that there are three basic errors in the quoted statement:

(i) the forces exerted by the road are ignored;

(ii) it does not distinguish between forces exerted *on* the horse and forces exerted *by* the horse;

(iii) action and reaction are simultaneous so 'first' is nonsense.

3.4 Some comments on Newton's laws of motion

Since Newton's laws were first proposed almost 300 years ago they have been found to be consistent with the results of an extremely wide range of experiments.

It was in 1905 that Einstein proposed, in the theory of Special Relativity, that Newton's laws needed to be modified for particles whose speeds are close to c, the speed of light. The value of c is about 3×10^8 ms^{-1}. The principal modification is embodied in the Second Law (3.2).

According to the theory of relativity, the mass m of a body is not constant but depends on the speed v with which the body is moving according to the formula

$$m = \frac{m_0}{\sqrt{\left(1 - \dfrac{v^2}{c^2}\right)}}, \tag{3.6}$$

where $v = |\mathbf{v}|$ is the speed of the body and m_0 is a constant equal to the body's mass when it is at rest (the *rest mass*).

We then define the *momentum* $\mathbf{p}$ of a particle by

$$\mathbf{p} = m\mathbf{v}, \tag{3.7}$$

where m is given in terms of m_0, v and c by (3.6). Thus $\mathbf{p}$ is a vector parallel to $\mathbf{v}$. Einstein's theory shows that Newton's Second Law must always be written in the form

$$\mathbf{F} = \dot{\mathbf{p}} = \frac{\mathrm{d}}{\mathrm{d}t}(m\mathbf{v}). \tag{3.8}$$

Since m varies when $\mathbf{v}$ varies, $\dot{m} \neq 0$. Hence, by the product rule of differentiation, the right-hand side of (3.8) is $m\dot{\mathbf{v}} + \dot{m}\mathbf{v}$. Since $\dot{\mathbf{v}} = \mathbf{a}$, the right-hand sides of (3.8) and (3.2) differ by the extra term $\dot{m}\mathbf{v}$ in (3.8). In fact Newton quoted his Second Law in the form (3.8). Therefore Einstein's original contribution is not (3.8) but equation (3.6) giving the dependence of m on m_0, v and c.

However the most important point is that in most everyday circumstances there is no *practical* difference between (3.2) and (3.8) because the speeds concerned are so much less than c that there are no observable differences between m and m_0. Suppose, for example, that we are considering the motion of an aircraft whose speed is 300 ms^{-1} (approximately the speed of sound). According to (3.6)

$$m = \frac{m_0}{\sqrt{\left(1 - \dfrac{1}{10^{12}}\right)}},$$

so that $m = m_0$ for all practical circumstances. Even for a particle moving at one tenth of the speed of light, m is only about $1{\cdot}005\, m_0$. In this book we shall be concerned, as explained earlier in §1.2, only with motions for which m and m_0 are effectively the same, so we can use either (3.2) or (3.8) and regard m as constant.

It is often pointed out that Newton's First Law (3.1) is the special case of Newton's Second Law (3.2) when $\mathbf{F} = \mathbf{0}$. This is of course true. However there is much to be said for keeping the two laws separate since stating the First Law is really equivalent to stating that it is possible to observe motions of particles at constant velocity when there are no forces acting on them. In other words, the First Law is really a statement that there exists a *frame of reference* (by which is meant an origin and a cartesian basis of three vectors) with the property that Newton's First Law (3.1) is true for an observer fixed relative to this frame. Such a frame is called an *inertial frame of reference.*

Exercise 3

1 A particle of mass m is moving so that at time t its position vector $\mathbf{r}$ with respect to a fixed origin is $\mathbf{r} = a(\mathbf{i}\cos 2\omega t + \mathbf{j}\sin 2\omega t)$, where a and ω are positive constants. Show that the force $\mathbf{F}$ acting on the particle is given by $\mathbf{F} = -4m\omega^2\mathbf{r}$. Calculate the magnitude of $\mathbf{F}$ when $a = 0{\cdot}5$ m, $\omega = 2\ \mathrm{s}^{-1}$ and $m = 0{\cdot}1$ kg.

2 At time t a ball of mass m has position vector $\mathbf{r}$ with respect to its initial position, where

$$\mathbf{r} = \mathbf{V}(1 - \mathrm{e}^{-kt})/k + \mathbf{g}(1 - kt - \mathrm{e}^{-kt})/k^2,$$

k being a constant positive scalar, and $\mathbf{V}$ and $\mathbf{g}$ being constant vectors. Determine the velocity $\mathbf{v}$ and acceleration $\mathbf{a}$ of the ball at time t. Hence show that the force $\mathbf{F}$ acting on the ball always satisfies $\mathbf{F} = -m(\mathbf{g} + k\mathbf{v})$. What are the units of k and $\mathbf{V}$?

3 According to a simple theory, the magnitude L of the lift force on an aeroplane of wing-span a is proportional to $a^m \rho^n v^p$, where ρ is the density of the air, v is the speed of the aeroplane, and m, n, p are numerical constants. Determine the values of m, n, p so that the theory is dimensionally consistent.

4 The magnitude W of the weight of a body of mass m is defined by $W = mg$, where g is a constant whose approximate value is 9·81 ms^{-2}. Show that 1 lb wt, the magnitude of the weight of a body of mass 1 lb, is approximately 4·45 N.

When a force of magnitude F acts upon an area A it is said to exert a *pressure* equal to (F/A). The air pressure in a car tyre is 28 psi (where 1 psi is the pressure exerted by a force of magnitude 1 lb wt acting on an area of 1 square inch). Determine this pressure in Nm^{-2}, giving your answer to three significant figures. Also show that an atmospheric pressure of 1000 millibars, where 1 millibar $= 10^2\ \mathrm{Nm}^{-2}$, is approximately 14·5 psi.

5 The rest mass of a proton is approximately $1{\cdot}67265 \times 10^{-27}$ kg. Estimate its mass when it is moving with speed $0{\cdot}995c$, where c is the speed of light, and also the magnitude of its momentum in $\mathrm{kg\,m\,s^{-1}}$.

4 Examples of particle motion

4.1 Introduction

In this chapter we describe some of the most common forces which cause particles to move, and then we use Newton's Second Law (3.2) to study some of the properties of these motions.

4.2 Motion under a constant force

Since we shall meet several important situations in which the force $\mathbf{F}$ acting on a particle is constant, it is useful first to derive and summarize some of the results that apply in all such cases. When $\mathbf{F}$ is constant, (3.2) shows that its effect on a particle of mass m is to give it a constant acceleration $\mathbf{a}$, where

$$\mathbf{a} = \frac{1}{m}\mathbf{F}. \tag{4.1}$$

Remember the term 'constant' means 'unchanging with time', and a constant vector is one whose magnitude and direction are both constant.

By (2.19) $\dot{\mathbf{v}} = \mathbf{a}$, where $\mathbf{v}(t)$ is the velocity of the particle at time t. Since $\mathbf{a}$ is constant, the derivative of $\mathbf{a}t$ is $\mathbf{a}$. Hence $\mathbf{v}$ and $\mathbf{a}t$ have the same derivative. In other words

$$\mathbf{v} = \mathbf{a}t + \mathbf{u}, \tag{4.2}$$

where $\mathbf{u}$ is a constant (vector) equal to the velocity of the particle when $t = 0$.

By (2.15) $\dot{\mathbf{r}} = \mathbf{v}$, where $\mathbf{r}(t)$ is the position vector of the particle with respect to the origin O at time t, and $\mathbf{v}$ is given in this case by (4.2). Since $\mathbf{a}$ and $\mathbf{u}$ are constant in (4.2), the derivative of $\frac{1}{2}\mathbf{a}t^2 + \mathbf{u}t$ is $\mathbf{a}t + \mathbf{u} = \mathbf{v}$ by (4.2). Hence $\mathbf{r}$ and $\frac{1}{2}\mathbf{a}t^2 + \mathbf{u}t$ differ only by a constant. Hence

$$\mathbf{r} = \tfrac{1}{2}\mathbf{a}t^2 + \mathbf{u}t + \mathbf{d}, \tag{4.3}$$

where the constant $\mathbf{d}$ is the position vector of the initial position of the particle. It is often convenient to choose the origin O to be this initial position; in that case $\mathbf{d} = \mathbf{0}$ in (4.3).

Example 1 A particle of mass 2 kg is acted on by a constant force $(4\mathbf{i} + 8\mathbf{j})$ N. At $t = 0$, the particle has velocity $(-\mathbf{j} + \mathbf{k})\ \mathrm{ms}^{-1}$ and position vector $(\mathbf{i} + \mathbf{k})$ m. Determine the position vector of the particle when $t = 1$ s, and find its distance from O at this time.

$$(4.1) \quad \Rightarrow \quad \mathbf{a} = (2\mathbf{i} + 4\mathbf{j})\ \mathrm{ms}^{-2}.$$

Putting $\mathbf{u} = (-\mathbf{j} + \mathbf{k})\ \mathrm{ms}^{-1}$, $\mathbf{d} = (\mathbf{i} + \mathbf{k})$ m, and $t = 1$ s in (4.3) gives

$$\mathbf{r} = [\tfrac{1}{2}(2\mathbf{i} + 4\mathbf{j}) + (-\mathbf{j} + \mathbf{k}) + (\mathbf{i} + \mathbf{k})]\ \mathrm{m}$$
$$= (2\mathbf{i} + \mathbf{j} + 2\mathbf{k})\ \mathrm{m}.$$

Hence the distance from O when $t = 1$ s is

$$|\mathbf{r}| = \sqrt{(2^2 + 1^2 + 2^2)}\ \mathrm{m} = 3\ \mathrm{m}.$$

Example 1 illustrates the use of (4.1) and (4.3) when the components of **a**, **u** and **d** are known.

In many applications the motion takes place along a straight line; that is **a** and **u** are in the same direction. We can then choose this direction to be **j** and put $\mathbf{a} = a\mathbf{j}$, $\mathbf{u} = u\mathbf{j}$. From (4.2), we obtain $\mathbf{v} = v\mathbf{j}$, and hence

$$v = u + at. \tag{4.4}$$

Equation (4.3) gives $(\mathbf{r} - \mathbf{d}) = (\frac{1}{2}at^2 + ut)\mathbf{j}$. This shows that the displacement of the particle relative to its initial position is also parallel to **j**. Let this displacement be $s\mathbf{j}$, so that

$$s = ut + \tfrac{1}{2}at^2. \tag{4.5}$$

There are many useful relationships between the quantities appearing in (4.4) and (4.5). For example, it follows from (4.4) that

$$v^2 = u^2 + 2uat + a^2t^2 = u^2 + 2a(ut + \tfrac{1}{2}at^2).$$

Hence

$$v^2 = u^2 + 2as, \tag{4.6}$$

where s is given in (4.5). Another relationship is

$$s = \tfrac{1}{2}(u + v)t. \tag{4.7}$$

The derivation of (4.7) is left as an exercise for the reader.

Examples of the use of equations (4.1)–(4.7) occur later in this chapter. It is however necessary to stress that all of the formulae except (4.1) apply only in situations where the force **F**, and hence the acceleration **a**, is constant.

4.3 Motion under gravity near the Earth's surface

One of the most important applications of the results in §4.2 is to motion under *gravity* near the Earth's surface. As explained briefly in §3.2, the Earth exerts a force on every particle. This force is directed vertically downwards, that is towards the centre of the Earth, and is called the *weight* of the particle. Let **j** be a unit vector vertically upwards. Then the weight **W** of a particle of mass m is given by the formula

$$\mathbf{W} = -W\mathbf{j}, \text{ where } W = mg. \qquad (4.8)$$

In (4.8), g is a quantity with the dimension of an acceleration. The value of g varies slightly from place to place on the Earth's surface. In England g is $9{\cdot}81$ ms^{-2} to a good approximation, and it can therefore be taken as 10 ms^{-2} in many rough calculations. This explains the choice of $g = 10$ ms^{-2} in the calculations for Fig. 2.1.

When a particle is above the Earth's surface, the magnitude of the force exerted on it by the Earth is less than mg. But, as we shall see in §4.4, the difference is insignificant unless the height is large, comparable indeed to the radius of the Earth, which is about $6{\cdot}4 \times 10^6$ m.

Therefore, in studying the motion of a particle near the Earth's surface, we can accept that the force of the Earth on the particle is constant (in magnitude and direction). Assume for the moment that the only force exerted on the particle is its weight **W** so that, in particular, we neglect any air resistance. By (4.1) and (4.8), the acceleration **a** of the particle is constant and equal to $-g\mathbf{j}$. Hence we can apply the formulae in §4.2. We begin with an example in which the motion is in a vertical straight line.

Example 2 A ball is thrown vertically upwards with an initial speed of $8{\cdot}4$ ms^{-1} from a point $1{\cdot}3$ m above the ground. Taking g as $9{\cdot}8$ ms^{-2}, and neglecting air resistance, find

(i) the maximum height above the ground reached by the ball;
(ii) the time before the ball hits the ground.

(i) We use (4.6) with $u = 8{\cdot}4$ ms^{-1} and $a = -9{\cdot}8$ ms^{-2}.

When the ball is at its maximum height, its speed v is zero. Let its height above the point of projection be s m.

$$(4.6) \quad \Rightarrow \quad 0 = (8{\cdot}4)^2 - 2 \times 9{\cdot}8 \times s \quad \Rightarrow \quad s = 3{\cdot}6.$$

$$\Rightarrow \quad \text{maximum height above ground} = (3{\cdot}6 + 1{\cdot}3) \text{ m} = 4{\cdot}9 \text{ m}.$$

(ii) The time t_1 s taken to reach maximum height can be obtained from (4.4) with $v = 0$, $u = 8{\cdot}4$ ms^{-1} and $a = -9{\cdot}8$ ms^{-2}.

$$(4.4) \quad \Rightarrow \quad 0 = 8{\cdot}4 - 9{\cdot}8t_1 \quad \Rightarrow \quad t_1 = (6/7).$$

The time t_2 s taken to fall from the maximum height to the ground $4{\cdot}9$ m below can be found from (4.5). If in this calculation we measure distance vertically downwards and time from the instant of achieving maximum height, we have $u = 0$, $a = 9{\cdot}8$ ms^{-2} and, from the result of (i), $s = 4{\cdot}9$ m.

$$(4.5) \quad \Rightarrow \quad 4{\cdot}9 = \tfrac{1}{2} \times 9{\cdot}8 \times t_2{}^2 \quad \Rightarrow \quad t_2 = 1.$$

$$\Rightarrow \quad \text{time before ball hits ground} = (t_1 + t_2) \text{ s} = (13/7) \text{ s}.$$

Alternatively, we can obtain the time T s before the ball hits the ground directly from (4.5) as follows. Measuring distance vertically upwards from the point of projection, we have $t = T$ s, $u = 8{\cdot}4$ ms^{-1}, $a = -9{\cdot}8$ ms^{-2} and $s = -1{\cdot}3$ m.

$$(4.5) \quad \Rightarrow \quad -1{\cdot}3 = 8{\cdot}4T - 4{\cdot}9T^2$$

$$\Rightarrow \quad 49T^2 - 84T - 13 = 0$$

$$\Rightarrow \quad (7T - 13)(7T + 1) = 0.$$

Thus $T = (13/7)$ as before, provided we reject the negative root $T = -(1/7)$. However, this negative root does have an interesting significance. Can you see what it is?

We now consider cases in which the motion is not in a straight line. This happens when the initial velocity $\mathbf{u}$ is not vertical. We have already chosen $\mathbf{j}$ to be vertically upwards; let us, in addition, choose the direction of $\mathbf{i}$ so that

$$\mathbf{u} = \mathbf{i}V\cos\alpha + \mathbf{j}V\sin\alpha. \tag{4.9}$$

In (4.9), $V > 0$ is the speed of projection, $\mathbf{i}$ is in a horizontal direction and, as Fig. 4.1 shows, α is the angle of projection above the horizontal. We can allow negative values of α; they arise when the vertical component of the initial velocity is towards the Earth as, for example, when a stone is thrown downwards from the top of a cliff.

We continue to take the acceleration equal to $-g\mathbf{j}$. The velocity $\mathbf{v}$ at time t after projection is given by (4.2) as

$$\mathbf{v} = \mathbf{i}V\cos\alpha + \mathbf{j}(V\sin\alpha - gt). \tag{4.10}$$

The position vector $\mathbf{r}$ of the particle at time t after projection is given by (4.3). For convenience we choose the origin O to be at the point of projection. Then $\mathbf{d} = \mathbf{0}$ in (4.3) so that

$$\mathbf{r} = \mathbf{i}Vt\cos\alpha + \mathbf{j}(Vt\sin\alpha - \tfrac{1}{2}gt^2). \tag{4.11}$$

Note that (4.10) and (4.11) are the same as (2.29) derived in Example 10, p. 22.

Equations (4.10) and (4.11) are the basic equations for the study of *projectiles*, which are particles moving in the atmosphere under the Earth's gravitational field when air resistance is neglected. We now develop and apply some of the principal results that can be obtained from (4.10) and (4.11). It is convenient to let x and y be the components of $\mathbf{r}$, so that from (4.11)

$$\mathbf{r} = x\mathbf{i} + y\mathbf{j}; \quad x = Vt\cos\alpha, \quad y = Vt\sin\alpha - \tfrac{1}{2}gt^2. \tag{4.12}$$

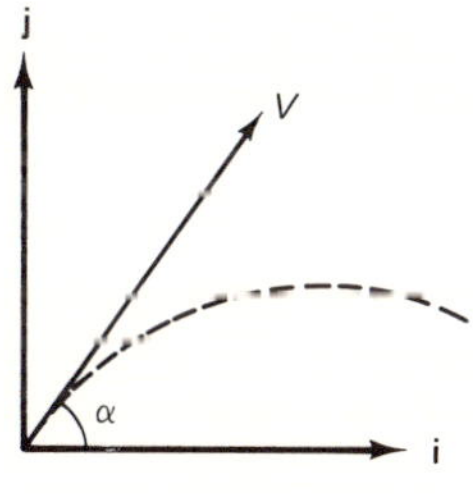

Fig. 4.1

Consider a projectile fired from a point on horizontal ground. It will hit the ground again when $y = 0$. From (4.12), this means that $0 = Vt \sin\alpha - \frac{1}{2}gt^2$. Thus $t = 0$, the instant of projection, when y is also zero, or $t = T$, where

$$T = \frac{2V\sin\alpha}{g}. \tag{4.13}$$

Thus T is the *time of flight*. The horizontal distance R travelled in time T is known as the *range* (on the horizontal plane through O). By substituting $t = T$ in the formula for x in (4.12), we find

$$R = \frac{2V^2\cos\alpha\sin\alpha}{g} = \frac{V^2\sin 2\alpha}{g}, \tag{4.14}$$

where we have used the trigonometric formula $\sin 2\alpha \equiv 2\sin\alpha\cos\alpha$. For a given speed of projection V, the range R has a maximum when $\sin 2\alpha$ has a maximum. This occurs when $2\alpha = \pi/2$. So the maximum range is V^2/g attained by firing in a direction 45° above the horizontal.

It is also of interest to determine the greatest height H above the ground of the projectile during its flight. The formula for y in (4.12) can be rearranged by completing the square to

$$y = \frac{V^2\sin^2\alpha}{2g} - \frac{1}{2}g\left(t - \frac{V\sin\alpha}{g}\right)^2.$$

Since $[t - (V\sin\alpha)/g]^2$ is zero when $t = (V\sin\alpha)/g$, and positive otherwise, y has its maximum value when $t = (V\sin\alpha)/g$, and substituting back into (4.12) gives

$$H = \frac{V^2\sin^2\alpha}{2g}. \tag{4.15}$$

This result may also be obtained from the formula for y in (4.12) by differentiating with respect to t and so obtaining $\dot{y} = V\sin\alpha - gt$. Stationary values of y occur when $\dot{y} = 0$; thus, as before, the maximum value of y is reached when $t = (V\sin\alpha)/g$ and substitution in the formula for y gives (4.15). Since $\dot{y}$ is the vertical component of $\mathbf{v}$, this method confirms the physically obvious result that the particle is at its maximum height above the ground when it has no vertical component of velocity, that is when it is instantaneously moving horizontally. Either method of derivation of (4.15) shows also that $y = H$ when $t = \frac{1}{2}T$, where T is the time of flight given by (4.13).

Example 3 A golfer strikes a golf ball so that its initial velocity makes an angle $\tan^{-1}(1/2)$ with the horizontal. Given that the range of the ball on the horizontal plane through the point of projection is 156·8 m, calculate the greatest height of the ball above the plane and its time of flight. Take g as 9·8 ms^{-2} and ignore air resistance.

Since $\alpha = \tan^{-1}(1/2)$,

$$\sin\alpha = 1/\sqrt{5}, \quad \cos\alpha = 2/\sqrt{5}, \quad \sin 2\alpha = 4/5.$$

Using (4.14) with $R = 156{\cdot}8$ m, $g = 9{\cdot}8$ ms^{-2}

$$\Rightarrow \quad V^2 = 1920{\cdot}8 \text{ m}^2\text{s}^{-2}.$$

$$(4.15) \quad \Rightarrow \quad H = (1920{\cdot}8 \times 0{\cdot}2)/(2 \times 9{\cdot}8) \text{ m} = 19{\cdot}6 \text{ m}$$

$$(4.13) \quad \Rightarrow \quad T = \sqrt{(1920{\cdot}8 \times 0{\cdot}2)/4{\cdot}9} \text{ s} = 4 \text{ s}.$$

In some circumstances it is necessary to know the equation of the path of the projectile. From the formula for x in (4.12), we obtain $t = (x \sec \alpha)/V$. Substituting in the formula for y in (4.12), we find

$$y = \frac{(V \sin \alpha)(x \sec \alpha)}{V} - \frac{gx^2 \sec^2 \alpha}{2V^2}$$

$$\Rightarrow \quad y = x \tan \alpha - \frac{gx^2}{2V^2} \sec^2 \alpha. \tag{4.16}$$

In using (4.16) (as in the following Example 4), it is often convenient to use the trigonometric identity $\sec^2 \alpha \equiv 1 + \tan^2 \alpha$ and rewrite (4.16) as

$$y = x \tan \alpha - \frac{gx^2}{2V^2}(1 + \tan^2 \alpha). \tag{4.17}$$

Equation (4.16)—or (4.17)—is the equation of a *parabola*; this can be seen most simply by rewriting the equation of the trajectory in the form

$$(y - H) = -\left(\frac{g \sec^2 \alpha}{2V^2}\right)(x - \tfrac{1}{2}R)^2,$$

where H and R are given in (4.15) and (4.14) respectively. When written in this form it is clear that the axis of the parabola is vertically downwards and that its vertex is at $x = \frac{1}{2}R$, $y = H$, that is the point where the projectile is at its maximum height above the point of projection.

Note also that the results in this section, though written in terms of the motion of projectiles, show that the path of *any* particle subject to *any* constant force (whatever its cause) is a parabola (exceptionally a straight line).

Example 4 A batsman strikes a ball at a height of 1 m above the ground giving it an initial speed of 25 ms^{-1} at an angle of 30° above the horizontal. The ball just clears the boundary without bouncing. Find the distance of the boundary from the batsman, neglecting air resistance and taking $g = 9{\cdot}8$ ms^{-2}.

Let the boundary be X m from the batsman's feet. Since x and y in (4.16) and (4.17) are measured from the point of projection, the boundary is at $x = X$ m, $y = -1$ m. Putting these values in (4.17) with $V = 25$ ms^{-1}, $g = 9{\cdot}8$ ms^{-2} and $\tan \alpha = 1/\sqrt{3}$ gives, after some simplification,

$$X^2 - 55{\cdot}2312X - 95{\cdot}6633 = 0.$$

This is a quadratic equation for X with two real roots, one positive and one negative. We obviously want the positive root. Use of the formula for the roots of a quadratic equation gives $X \approx 56{\cdot}9$.

We conclude this section with two further worked examples illustrating other types of problem that can be solved using the methods described above.

Example 5 A shell is fired with muzzle speed V from a gun situated at a point A on the sea shore to hit a vessel which, at the instant of firing, is at a point B and moving with constant speed $U(<V)$ in a direction perpendicular to AB. If α is the angle of elevation of the gun and $AB = d$, prove that

$$g^2d^2 = 4V^2\sin^2\alpha(V^2\cos^2\alpha - U^2). \tag{4.18}$$

Show that the vessel cannot be hit if $gd > V^2 - U^2$.

For this problem a diagram is essential. As Fig. 4.2 shows, the shell must not be aimed to land at B but at C, another point on the vessel's track, at the same time as the vessel reaches C. Therefore, if T denotes the time of flight, the point C must be chosen so that $BC = UT$. Also $AC = R$, the range of the shell.

By Pythagoras' Theorem

$$AC^2 = AB^2 + BC^2 \quad \Rightarrow \quad R^2 = d^2 + U^2T^2.$$

$$(4.14) \quad \Rightarrow \quad gR = 2V^2\cos\alpha\sin\alpha;$$

$$(4.13) \quad \Rightarrow \quad gT = 2V\sin\alpha.$$

$$\Rightarrow \quad g^2d^2 = g^2R^2 - g^2U^2T^2 = 4V^4\cos^2\alpha\sin^2\alpha - 4V^2U^2\sin^2\alpha.$$

$$\Rightarrow \quad g^2d^2 = 4V^2\sin^2\alpha(V^2\cos^2\alpha - U^2).$$

Use of $\cos^2\alpha \equiv 1 - \sin^2\alpha \quad \Rightarrow \quad g^2d^2 = 4V^2\sin^2\alpha[(V^2 - U^2) - V^2\sin^2\alpha]$

$$\Rightarrow \quad g^2d^2 = (V^2 - U^2)^2 - [2V^2\sin^2\alpha - (V^2 - U^2)]^2.$$

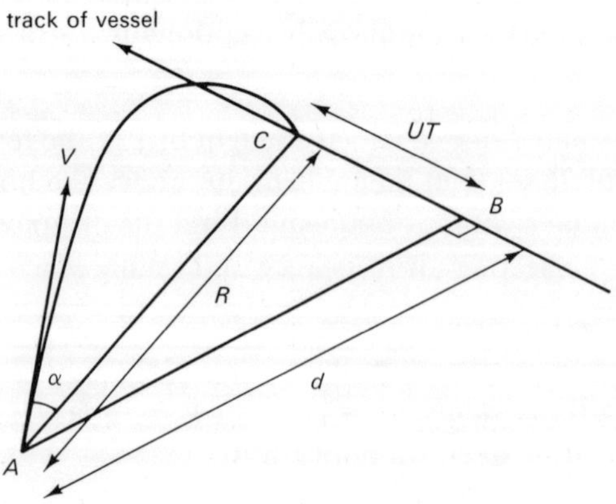

Fig. 4.2 Sketch illustrating Example 5.

Since a square is never negative, it follows that (4.18) cannot be satisfied for any α if $g^2d^2 > (V^2 - U^2)^2$; that is the vessel cannot be hit if $gd > V^2 - U^2$. (Note that when $gd \leqslant V^2 - U^2$ the vessel can be hit, and (4.18) is an equation from which the value of α that ensures a hit can be determined.)

Example 6 A shell bursts on level ground throwing fragments with initial speed V in all directions. After a time T a fragment hits the ground at a distance R from the place where the shell burst. Show that

$$g^2T^4 - 4V^2T^2 + 4R^2 = 0.$$

Hence, taking g as 10 ms^{-2}, estimate the period of time during which a man standing 30 m from the shell is in danger of suffering a direct hit when $V = 30$ ms^{-1}.

$$\begin{aligned}
(4.13) \quad &\Rightarrow \quad gT = 2V \sin\alpha \\
&\Rightarrow \quad 4V^2 \sin^2\alpha = g^2T^2, \\
&\qquad\;\; 4V^2 \cos^2\alpha = 4V^2 - g^2T^2. \\
(4.14) \quad &\Rightarrow \quad 4R^2 = (4V^2 \sin^2\alpha)(4V^2\cos^2\alpha)/g^2 \\
&\qquad\qquad\;\; = 4V^2T^2 - g^2T^4,
\end{aligned}$$

on substituting for $\sin^2\alpha$ and $\cos^2\alpha$,

$$\Rightarrow \quad g^2T^4 - 4V^2T^2 + 4R^2 = 0.$$

This is a quadratic equation for T^2 whose roots are

$$T^2 = \left(\frac{2V^2}{g^2}\right)\left[1 \pm \left(1 - \frac{g^2R^2}{V^4}\right)^{1/2}\right],$$

$$\Rightarrow \quad T^2 = (18 \pm 12\sqrt{2})\ \text{s}^2 = 6(\sqrt{2} \pm 1)^2\ \text{s}^2$$

on substituting the given values of g, R and V. The two positive values of T are therefore $\sqrt{6}(\sqrt{2} + 1)$ s and $\sqrt{6}(\sqrt{2} - 1)$ s. The man is in danger between these two times, that is for a period of $2\sqrt{6}$ s $\approx$ 4·9 s.

4.4 Newton's law of gravity

According to Newton's law of gravity every particle in the universe exerts a force on every other particle. Consider two particles of mass M and m located at A and B respectively; the situation is illustrated in Fig. 4.3. Let the directed line segment $\vec{AB}$ specify the vector $\mathbf{r}$, so that $AB = |\mathbf{r}| = r$.

The law of gravity states that the particle of mass M at A exerts a force $\mathbf{F}$ on the particle of mass m at B. The direction of $\mathbf{F}$ is *towards* A, that is parallel to $-\mathbf{r}$, and the magnitude of $\mathbf{F}$ is GMm/r^2, where G is Newton's *gravitational constant* equal to $6{\cdot}67 \times 10^{-11}$ m^{3}kg^{-1}s^{-2}. Since $(\mathbf{r}/r)$ is a unit vector parallel to $\mathbf{r}$, the force $\mathbf{F}$ exerted *by* the particle of mass M *on* the particle of mass m is therefore given by

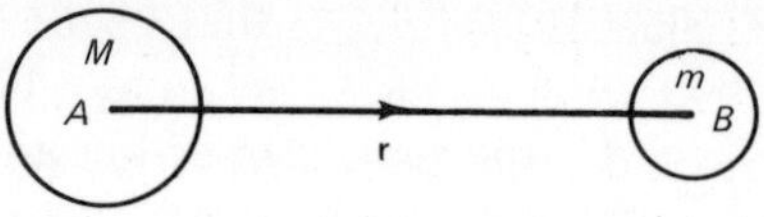

Fig. 4.3

$$\mathbf{F} = -\frac{GMm}{r^2}\frac{\mathbf{r}}{r} = -\frac{GMm}{r^3}\mathbf{r}. \tag{4.19}$$

(Note that **F** is an *attractive* force; that is it acts on the particle of mass m in the direction that tends to reduce $AB = r$.)

By Newton's Third Law (3.3), the particle of mass m exerts a force $-\mathbf{F}$ on the particle of mass M.

Although the law of gravity was formulated in terms of particles, it can be shown to have much wider application. For example the gravitational force between two uniform *spherical* bodies of mass M and m with centres A and B respectively is also given by (4.19), and the gravitational force between two bodies of any shape is also given to good approximation by (4.19) provided the distance between the bodies is much greater than their average diameters.

An example illustrating the use of (4.19) has already been given in Example 4, p. 12, and, as shown by Newton and others, (4.19) gives the force that causes planets to move around the Sun in closed orbits.

We wish to reconcile (4.19) with the result of (4.8), namely that the Earth exerts a force of magnitude mg on a body of mass m at its surface, and that this force is vertically downwards. Since 'vertically downwards' means 'towards the centre of the Earth', (4.19) gives the observed direction. It remains only to demonstrate that the magnitude of **F** in (4.19) is mg. But, from (4.19), $|\mathbf{F}| = GMm/r_E{}^2$, where r_E is the radius of the Earth and M is its mass. Using the values given in Example 4, p. 12, we find, as expected,

$$\frac{1}{m}|\mathbf{F}| = \left[\frac{6{\cdot}67 \times 10^{-11} \times 5{\cdot}98 \times 10^{24}}{(6{\cdot}37 \times 10^6)^2}\right] \text{ms}^{-2} \approx 9{\cdot}8 \text{ ms}^{-2}.$$

The conclusion is that (4.8) and (4.19) are consistent, and that

$$g = \frac{GM}{r_E{}^2}. \tag{4.20}$$

Now consider a body of mass m at a height h above the Earth's surface, so that its distance from the centre of the Earth is $(r_E + h)$. By (4.19), the Earth exerts a force of magnitude $|\mathbf{F}|$ on the body, where

$$|\mathbf{F}| = \frac{GMm}{(r_E + h)^2} = \frac{GMm}{r_E{}^2}\left(1 + \frac{h}{r_E}\right)^{-2}$$

and using (4.20)

$$= mg\left(1 + \frac{h}{r_E}\right)^{-2}. \tag{4.21}$$

In §4.3, we assumed that $|\mathbf{F}|$ is equal to mg, and we now see that our assumption is justified provided $(1 + h/r_E)^{-2}$ can be approximated by 1. Since $r_E \approx 6{\cdot}37 \times 10^6$ m, this is true unless h is very large. For example $(1 + h/r_E)^{-2}$ does not fall below 0·95 until h is greater than about $1{\cdot}66 \times 10^5$ m.

4.5 Air resistance

In §4.3 we neglected variations in the gravitational force on a particle as it moved near the Earth's surface, and this is shown to be justified in §4.4. But we also neglected all other forces on the particle, particularly air resistance; this neglect is not usually justified.

When a body moves in a *fluid*, which means a gas or a liquid, it pushes the particles of fluid which therefore accelerate. By Newton's Second Law applied to the fluid, the body must exert a force on the fluid. By Newton's Third Law (3.3), the fluid exerts an equal and opposite force on the body. The component of this force in the opposite direction to that of the motion of the body is known as *drag*, or, when the fluid is air, as *air resistance*.

The drag always opposes the motion of the body. Its magnitude depends in a very complicated way on both the properties of the fluid and the way in which the body is moving, especially its speed v, where $v = |\mathbf{v}|$. For *very low speeds*, the magnitude of the drag is proportional to v and, as a vector, the drag on the body is $-mk\mathbf{v}$, where k is a positive constant. For a body also acted on by gravity, the total force $\mathbf{F}$ on the body is therefore $-mg\mathbf{j} - mk\mathbf{v}$, where $\mathbf{j}$ is a unit vector pointing vertically upwards. By Newton's Second Law (3.2), it then follows that the equation of motion of the body is

$$m\mathbf{a} = -mg\mathbf{j} - mk\mathbf{v}$$

$$\Rightarrow \quad \frac{d\mathbf{v}}{dt} + k\mathbf{v} = -g\mathbf{j}, \tag{4.22}$$

since, according to (2.19), $\mathbf{a}$ is the rate of change of $\mathbf{v}$ with respect to time.

However (4.22) only applies for very low speeds. Normally, the magnitude of the drag is proportional to v^α, where $\alpha > 1$, and a good approximation in practice is that it is proportional to v^2. As a vector, the drag on the body can therefore be written $-mkv^2\hat{\mathbf{v}}$, where k is a positive constant and $\hat{\mathbf{v}}$ is a unit vector parallel to $\mathbf{v}$. Thus $\hat{\mathbf{v}} = \mathbf{v}/|\mathbf{v}|$ and, since $v^2 = |\mathbf{v}|^2$, the drag is also equal to $-mk|\mathbf{v}|\mathbf{v}$. The equation of motion analogous to (4.22) is now

$$\frac{dv}{dt} + k|\mathbf{v}|\mathbf{v} = -g\mathbf{j}. \tag{4.23}$$

Note that the acceleration $\frac{dv}{dt}$ in both (4.22) and (4.23) depends on $\mathbf{v}$, and is therefore not constant. Consequently the results derived in §4.2 cannot be used. It is beyond the scope of this book to consider the effect of air resistance on projectiles in general, and we shall confine our attention to the special case of a particle moving in a vertical straight line. The use of (4.22) in such cases is typified by Example 5 in §2.2 (p. 13) for, if we have a particle moving vertically downwards with speed v, we can write $\mathbf{v} = -v\mathbf{j}$ in (4.22) and obtain

$$\frac{dv}{dt} = g - kv,$$

which was the equation solved in Example 5, p. 13. Therefore we now concentrate on cases where the air resistance is proportional to v^2.

Example 7 A ball of mass m is thrown vertically upwards with initial speed V. Its speed vertically upwards at time t after release is v, and the air resistance has magnitude mkv^2, where k is a positive constant. Show that the time taken for the ball to reach its maximum height is T, where

$$T = \frac{1}{\sqrt{(gk)}} \arctan\left[\sqrt{\left(\frac{kV^2}{g}\right)}\right]. \tag{4.24}$$

Given that $V = 20\ \text{ms}^{-1}$, $g = 9{\cdot}8\ \text{ms}^{-2}$ and $k = 0{\cdot}005\ \text{m}^{-1}$, show that T is about 6% less than it would be with no air resistance.

Since the ball is travelling upwards, we put $\mathbf{v} = v\mathbf{j}$ in (4.23). Also $|\mathbf{v}| = v$, so (4.23) gives

$$\frac{dv}{dt} = -(g + kv^2) \tag{4.25}$$

$$\Rightarrow \quad \frac{1}{(g + kv^2)} \frac{dv}{dt} = -1$$

$$\Rightarrow \quad \frac{d}{dt}\left\{\frac{1}{\sqrt{(gk)}} \arctan\left[\sqrt{\left(\frac{kv^2}{g}\right)}\right]\right\} = \frac{d}{dt}(-t).$$

Here we have used the result

$$\frac{d}{dt}\left[\arctan\left(\frac{t}{a}\right)\right] = \frac{a}{a^2 + t^2},$$

with the constant a equal to $\sqrt{(g/k)}$. Using the same idea as in Example 5, p. 13, gives

$$\frac{1}{\sqrt{(gk)}} \arctan\left[\sqrt{\left(\frac{kv^2}{g}\right)}\right] = C - t, \tag{4.26}$$

where C is a constant. When $t = 0$, $v = V$ so that

$$C = \frac{1}{\sqrt{(gk)}} \arctan\left[\sqrt{\left(\frac{kV^2}{g}\right)}\right].$$

When the particle is at its greatest height $v = 0$ and $t = T$. Since $\arctan 0 = 0$, we find $T = C$ as required.

Without air resistance, let the time to maximum height be T_0. The acceleration is now constant and equal to $-g\mathbf{j}$. We put $v = 0$, $u = V$ and $\mathbf{a} = -g$ in (4.4) and obtain $T_0 = V/g$. Hence

$$\frac{T}{T_0} = \frac{\arctan\left[\sqrt{\left(\dfrac{kV^2}{g}\right)}\right]}{\sqrt{\left(\dfrac{kV^2}{g}\right)}}.$$

With the given numerical values $\sqrt{(kV^2/g)} \approx 0{\cdot}452$. Using a calculator, $T/T_0 \approx (0{\cdot}425/0{\cdot}452) \approx 0{\cdot}94$, that is T is about 6% less than T_0.

To find the value of the maximum height H reached by the ball in this last example (Example 7), we use (2.4) to rewrite (4.25) as

$$v\frac{\mathrm{d}v}{\mathrm{d}x} = -(g + kv^2). \tag{4.27}$$

$$\Rightarrow \quad \left(\frac{v}{g + kv^2}\right)\frac{\mathrm{d}v}{\mathrm{d}x} = -1$$

$$\Rightarrow \quad \frac{\mathrm{d}}{\mathrm{d}x}\left[\frac{\ln(g + kv^2)}{2k}\right] = \frac{\mathrm{d}}{\mathrm{d}x}(-x).$$

Proceeding as in Example 7, we obtain

$$\ln(g + kv^2) = \ln(g + kV^2) - 2kx$$

$$\Rightarrow \quad \ln g = \ln(g + kV^2) - 2kH$$

$$\Rightarrow \quad H = \frac{1}{2k}\ln\left(1 + \frac{kV^2}{g}\right). \tag{4.28}$$

Without air resistance the maximum height reached H_0 can be found from (4.6), and is $V^2/(2g)$

$$\Rightarrow \quad \frac{H}{H_0} = \frac{\ln\left(1 + \dfrac{kV^2}{g}\right)}{\left(\dfrac{kV^2}{g}\right)},$$

and is about 0·91 with the numerical values of Example 7.

Once the ball reaches its maximum height, it begins to fall. Let its speed subsequently be v so that $v = -\mathbf{v}\mathbf{j}$, $|\mathbf{v}| = v$, remembering that $\mathbf{j}$ is vertically upwards. Then (4.23) leads to the equations

$$\frac{\mathrm{d}v}{\mathrm{d}t} = (g - kv^2) \quad \text{or} \quad v\frac{\mathrm{d}v}{\mathrm{d}x} = (g - kv^2), \tag{4.29}$$

which replace (4.25) and (4.27) respectively in the downwards motion. Notice that in the upwards motion both gravity and air resistance act against the motion, but in the downwards motion gravity acts to increase the speed whereas air resistance still acts to decrease it. These differences explain the different arrangements of signs in the equations for the different parts of the motion. Equation (4.29) is needed in questions 13, 14 and 15 in Exercise 4.

4.6 Reaction between surfaces, and friction

We now consider another important class of problems—those in which a moving body remains in contact with a fixed surface throughout the motion. For simplicity we shall suppose the fixed surface to be a plane. However we shall not suppose that the surface is necessarily horizontal.

As a first example consider the situation shown in Fig. 4.4 in which a body of mass m is moving in contact with a horizontal plane surface. The forces acting on the body are shown on the figure. First, there is a force $P\mathbf{i}$ causing motion; this could result from, for example, pushing the body, or pulling it with a rope. Secondly, there is the weight $-mg\mathbf{j}$ of the body. Finally, there is the force $\mathbf{R}$ exerted by the surface on the body. This force is called the *reaction* of the surface on the body. Experience shows that these three forces are the only ones influencing the motion. Then, denoting the acceleration of the body by $a\mathbf{i}$, where a may be positive or negative, Newton's Second Law (3.2) gives

$$P\mathbf{i} - mg\mathbf{j} + \mathbf{R} = ma\mathbf{i}. \tag{4.30}$$

Now let us write $\mathbf{R}$ in terms of its components with respect to the basis $\{\mathbf{i}, \mathbf{j}, \mathbf{k}\}$, where $\mathbf{i}$ and $\mathbf{j}$ are shown in Fig. 4.4, and $\mathbf{k}$ points out of the page. Every term except $\mathbf{R}$ in (4.30) is either parallel to $\mathbf{i}$ or parallel to $\mathbf{j}$; hence $\mathbf{R}$ can have no component in the direction of $\mathbf{k}$. Experience shows that the component of $\mathbf{R}$ normal to the surface is away from the surface, and the component of $\mathbf{R}$ along the surface opposes the motion. Thus

$$\mathbf{R} = -F\mathbf{i} + N\mathbf{j}, \tag{4.31}$$

where $F(=|\mathbf{R}|\sin\theta)$ and $N(=|\mathbf{R}|\cos\theta)$ are positive. It is conventional to call N the *normal reaction* and F the *frictional force* (or simply the *friction*). Substitution of (4.31) into (4.30) gives

$$(P - F)\mathbf{i} + (N - mg)\mathbf{j} = ma\mathbf{i}$$

$$\Rightarrow \quad P - F = ma, \quad N - mg = 0, \tag{4.32}$$

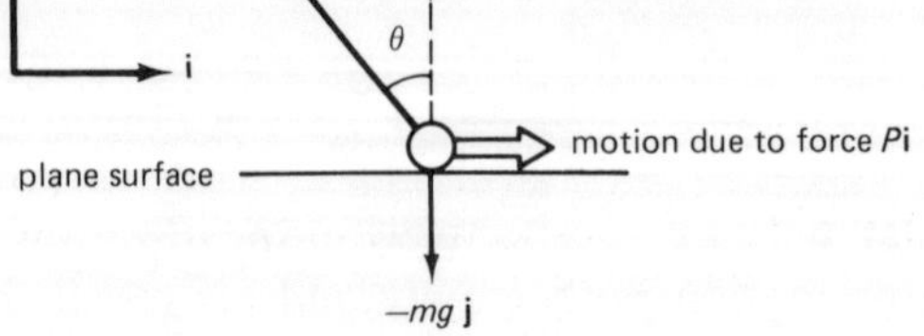

Fig. 4.4

where the last two equations arise by equating the coefficients of **i** and **j**. Therefore the normal reaction is equal to the magnitude of the weight of the body; if this were not so, the body would have a component of acceleration normal to the surface—contrary to the assumption that the body is moving on the surface. Now consider the equation $P - F = ma$. We can, of course, choose P by giving the body as much push (or pull) as we wish, and we know m. Before we can determine the acceleration a, we need to know F. (Many of the ideas in this paragraph were met earlier in Example 3, p. 31.)

Experiments show that when two surfaces are in relative motion (here the two surfaces are the horizontal plane and the surface of the body), the value of F is a constant μ times the value of N; that is

$$F = \mu N. \tag{4.33}$$

The constant μ is known as the *coefficient of friction*. Its value depends in detail on the nature of the surfaces in contact but it is invariably less than 1. In some circumstances the value of μ is so low that it is permissible to take $\mu = 0$ as a first approximation. Surfaces for which $\mu = 0$ are said to be *smooth*, and surfaces for which $\mu \neq 0$ are said to be rough.

It should be stressed that (4.33) is true only when the two surfaces are in relative motion.

From (4.32) and (4.33) it follows that, for the situation shown in Fig. 4.4, $F = \mu mg$ and $a = (P/m) - \mu g$.

Example 8 A body is sliding under its own weight down a line of greatest slope of a rough plane inclined at an angle α to the horizontal. The coefficient of friction is μ. Find the acceleration of the body, and deduce that the motion maintains itself only if $\tan\alpha \geqslant \mu$.

The situation is illustrated in Fig. 4.5. With **i** and **j** as shown, the forces acting on the body are its weight $mg(\sin\alpha\mathbf{i} - \cos\alpha\mathbf{j})$ and the reaction of the plane $(-F\mathbf{i} + N\mathbf{j})$. By Newton's Second Law,

$$(mg\sin\alpha - F)\mathbf{i} + (N - mg\cos\alpha)\mathbf{j} = ma\mathbf{i}$$

$$\Rightarrow \quad mg\sin\alpha - F = ma, \quad N = mg\cos\alpha.$$

$$(4.33) \quad \Rightarrow \quad F = \mu N \quad \Rightarrow \quad F = \mu mg\cos\alpha$$

$$\Rightarrow \quad a = g(\sin\alpha - \mu\cos\alpha).$$

If $a < 0$ the speed of the body down the plane decreases, eventually to zero. Therefore the motion is maintained only if $a \geqslant 0$

$$\Rightarrow \quad \sin\alpha - \mu\cos\alpha \geqslant 0$$

$$\Rightarrow \quad \tan\alpha \geqslant \mu.$$

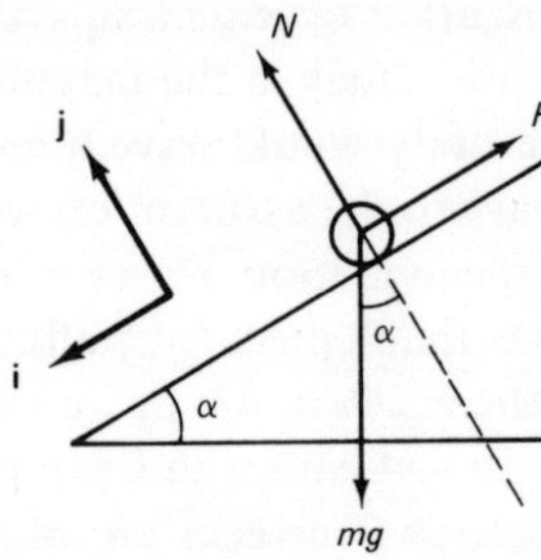

Fig. 4.5 Sketch illustrating Example 8.

4.7 Forces in taut strings and rigid rods

In discussing the situation shown in Fig. 4.4, p. 48, we noted that one way of causing the motion of the body was to pull it with a rope. Also, the traces joining the horse and cart in Example 3, p. 31, were the means of exerting a force of magnitude T on the cart and an equal and opposite force on the horse. There are many similar examples where forces are transmitted from one body to another through strings or rods. In this section we shall suppose that all the strings or rods considered are *light*, which means that their masses can be neglected.

As shown in Fig. 4.6, consider a small element PQ of a light string or rod. Let the part of the string or rod to the left of P exert a force of magnitude S on PQ in the direction away from PQ, and let the part to the right of Q exert a force of magnitude T on PQ in the direction away from PQ. Apply Newton's Second Law (3.2) to PQ. Since the string or rod is light, the mass of PQ is zero. Hence the total force on PQ is zero, so that $S = T$. This argument can be applied to every element of the string or rod, and to the sum of all elements, that is to the whole string or rod AB. Thus a light string or rod transmits an unchanged force from one end to the other.

The difference between a string and a rigid rod is that a rigid rod can transmit any force, whether it tends to stretch it or compress it. When the force tends to stretch the rod, it is said to be under *tension*; when the force tends to compress the rod, it is said to be under *thrust*. On the other hand, a light string can only transmit a force when it is under tension. A string cannot resist a force which tends to shorten it—it just collapses!

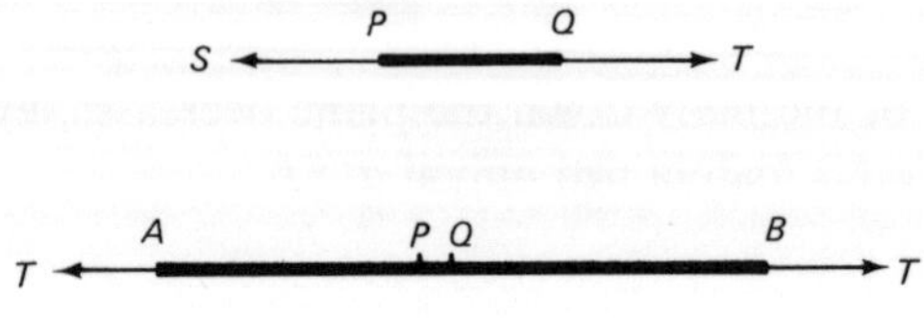

Fig. 4.6

In the remainder of this section we shall consider only strings and rods whose lengths do not change when they transmit forces; such strings and rods are called *inextensible* or *inelastic*. Actually, the lengths of all real strings and rods change (slightly at least) when they are subject to forces, and our assumption means that we shall consider only situations in which these changes have no significant effect. In §4.8 we shall consider elastic strings, in which changes in length are important.

Example 9 Two particles of mass $2m$ and m are joined by a light inextensible string which passes over a light, smooth pulley. The system is held at rest, with both parts of the string taut and vertical, and then released. Find the acceleration of each particle and the magnitude of the tension in the string.

The situation is illustrated in Fig. 4.7. The term 'light, smooth pulley' means that the pulley has no mass and that its bearing is smooth. Hence no force is required to turn the pulley and, as explained above, the tension in the string is equal throughout its length. Let the magnitude of the tension be T. The string will remain taut after release so that if the acceleration of the particle of mass $2m$ is of magnitude a and downwards, then the acceleration of the particle of mass m is also of magnitude a but upwards. Apply Newton's Second Law (3.2) to each particle in turn. Only the vertical component of (3.2) is non-trivial and we obtain

$$2mg - T = 2ma; \quad T - mg = ma.$$

These are a pair of simultaneous equations for a and T with solution

$$a = \tfrac{1}{3}g, \quad T = \tfrac{4}{3}mg.$$

The next example discusses a situation in which a light string provides the force necessary for a particle to move in a circle.

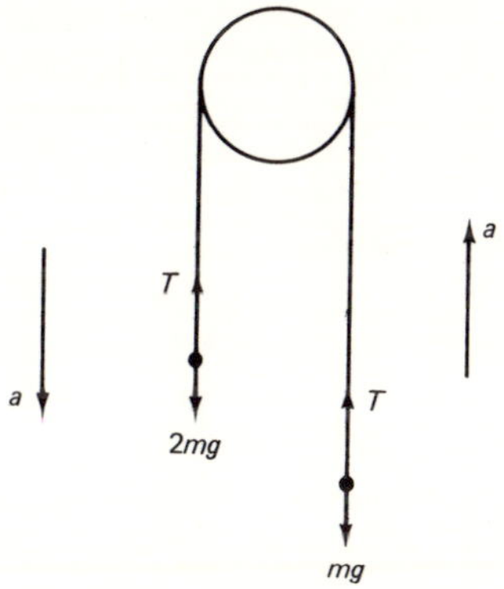

Fig. 4.7 Sketch illustrating Example 9.

Example 10 A small smooth ring P is threaded on a light inextensible string of length 1·4 m. The two ends of the string are fastened to two fixed smooth bearings A and B, where A is 1 m vertically above B. The ring is moving in a horizontal circle with constant angular speed so that $\angle APB = 90°$. Show that the radius of the circle is 0·48 m and that the magnitude of the tension in the string is five times the magnitude of the weight of the ring. Taking $g = 9{\cdot}81\ \text{ms}^{-2}$, find the angular speed of the ring in radians per second to 1 decimal place.

Once more a diagram is essential; the situation is illustrated in Fig. 4.8. Since the ring P is smooth, it exerts no frictional force on the string, and since the string is light, it follows that the magnitude T of the tension in the string is the same throughout the string. Take unit vectors $\mathbf{i}$ and $\mathbf{j}$ as shown, and let $\angle NAP = \alpha$. Denote the angular speed of P by ω. By (2.23), the acceleration of P is $-(\omega^2 PN)\mathbf{i}$.

The forces on P are (i) its weight $-W\mathbf{j}$, (ii) the tension due to the part AP of the string $-T(\sin\alpha\mathbf{i} - \cos\alpha\mathbf{j})$, (iii) the tension due to the part BP of the string $-T(\cos\alpha\mathbf{i} + \sin\alpha\mathbf{j})$. Since the mass of P is W/g (see (4.8)), Newton's Second Law gives

$$(-T\cos\alpha - T\sin\alpha)\mathbf{i} + (T\cos\alpha - T\sin\alpha - W)\mathbf{j} = -(W\omega^2 PN/g)\mathbf{i}$$

$$\Rightarrow \quad T(\cos\alpha + \sin\alpha) = (W\omega^2 PN/g), \quad T(\cos\alpha - \sin\alpha) = W.$$

We now have to find α and PN. Let the length of BP be x m. Since $APB = 90°$, using Pythagoras' Theorem

$$\Rightarrow \quad x^2 + (\tfrac{7}{5} - x)^2 = 1 \quad \Rightarrow \quad x = \tfrac{3}{5} \text{ or } \tfrac{4}{5}.$$

The equation $\quad T(\cos\alpha - \sin\alpha) = W \quad \Rightarrow \quad \cos\alpha > \sin\alpha \quad \Rightarrow \quad AP > BP$

$$\Rightarrow \quad AP = (4/5)\text{ m}, \quad BP = (3/5)\text{ m}.$$

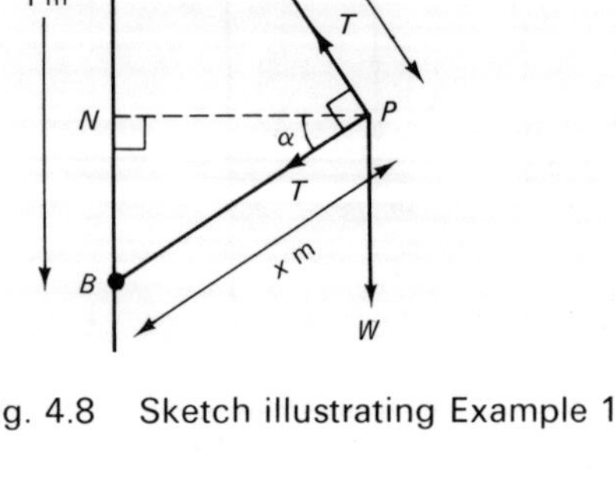

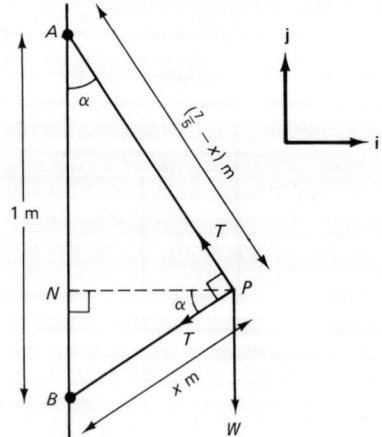

Fig. 4.8 Sketch illustrating Example 10.

Let r m be the length of PN.
Since $\sin\alpha = PN/AP = 5r/4$, $\cos\alpha = PN/BP = 5r/3$.

$$\sin^2\alpha + \cos^2\alpha = 1 \quad\Rightarrow\quad r = (12/25) \quad\Rightarrow\quad \text{length of } PN \text{ is } 0{\cdot}48 \text{ m.}$$

$$\cos\alpha - \sin\alpha = (4/5) - (3/5) = (1/5) \quad\Rightarrow\quad T(1/5) = W \quad\Rightarrow\quad T = 5W$$

$$\cos\alpha + \sin\alpha = (7/5) \quad\Rightarrow\quad \omega^2/g = (7/0{\cdot}48) \text{ m}$$

$$g = 9{\cdot}81 \text{ ms}^{-2} \quad\Rightarrow\quad \omega^2 \approx 143{\cdot}06 \text{ s}^2 \quad\Rightarrow\quad \omega \approx 12{\cdot}0 \text{ s.}$$

4.8 Springs and elastic strings

It is common experience that strings made of rubber and other similar materials can be stretched very easily, and that, when they are stretched, they exert a force on whatever agency or body is doing the stretching. The direction of this force is that which tends to restore the string to its original unstretched length. For this reason the force is sometimes called a *restoring force*, and strings in which restoring forces are set up by stretching are called *elastic strings*.

Let the original unstretched length of an elastic string be l; the usual term for l is the *natural length* of the string. Suppose the string is stretched to a length $(l + x)$ where x is positive and is known as the *extension* of the string. Experiments show that, provided (x/l) is not too large, the magnitude of the restoring force is proportional to x. This is *Hooke's Law*.

Hooke's Law is also valid for restoring forces set up in *springs* with the difference, from elastic strings, that a restoring force exists in a spring not only when it is stretched but also when it is compressed.

Consider the situation shown in Fig. 4.9(i). A particle P is attached to one end of an elastic string of natural length l; the other end is fastened to a fixed support at A. When the string is stretched so that its length is $(l + x)$, where $x > 0$, Hooke's Law shows that there is a positive constant k such that the

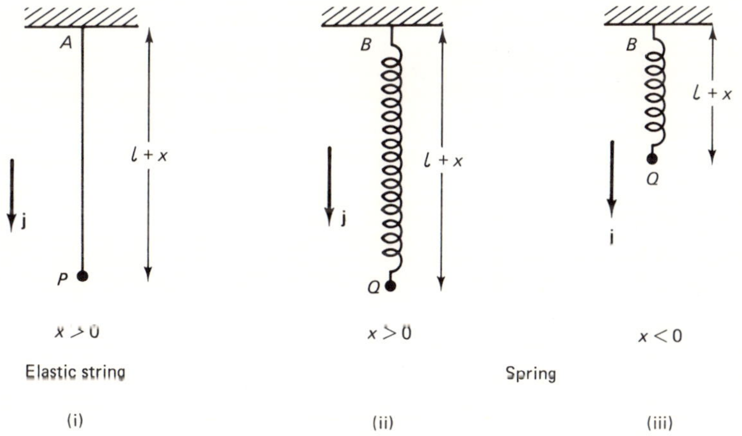

Fig. 4.9

magnitude $|\mathbf{T}|$ of the restoring force $\mathbf{T}$ on p is equal to kx. Let $\mathbf{j}$ be a unit vector vertically downwards; then $\mathbf{T}$ is in the opposite direction to $\mathbf{j}$. Hence

$$\mathbf{T} = -kx\mathbf{j}. \tag{4.34}$$

Figures 4.9(ii) and (iii) show a particle Q attached to one end of a spring of natural length l, the other end of which is fastened to a fixed support at B. Equation (4.34) obviously gives the restoring force when the spring is stretched to a length $(l + x)$, where $x > 0$. It also gives $\mathbf{T}$ when the spring is compressed to a length $(l + x)$, where $x < 0$, because the magnitude $|\mathbf{T}|$ of $\mathbf{T}$ is $k|x|$, as predicted by Hooke's Law, and, since $x < 0$, the direction of $\mathbf{T}$ is vertically downwards, as is necessary for $\mathbf{T}$ to be a force restoring the spring to its natural length.

The constant k is known as the *stiffness* of the spring. It is very common to express k in terms of l and another constant λ, called the modulus of elasticity. The equation relating k, λ and l is $k = \lambda/l$. Thus (4.34) can be written

$$\mathbf{T} = -T\mathbf{j} \quad \text{where } T = \frac{\lambda x}{l}. \tag{4.35}$$

The dimension of $\mathbf{T}$ is MLT^{-2}, since it is a force. So, from (4.35), the dimension of λ is also MLT^{-2}. The value of λ is a property of the spring.

We now begin a study of the type of motion caused by a restoring force $\mathbf{T}$ given by (4.34) or (4.35). Suppose that each of the particles P and Q in Fig. 4.9 has mass m so that, in addition to $\mathbf{T}$, each is also subjected to its weight $mg\mathbf{j}$. The net force on each particle is therefore $mg\mathbf{j} + \mathbf{T}$, and since the acceleration of each particle is $\ddot{x}\mathbf{j}$, Newton's Second Law (3.2) gives

$$m\ddot{x}\mathbf{j} = mg\mathbf{j} + \mathbf{T}.$$

Provided $\mathbf{T}$ is given by (4.35), we then find

$$m\ddot{x} = mg - (\lambda x/l). \tag{4.36}$$

Equation (4.36) is the equation of motion for Q for all values of x, and the equation of motion for P provided $x > 0$. It is convenient to rewrite (4.36) in the form

$$\ddot{x} + n^2 x = g, \quad n^2 = \frac{\lambda}{ml}. \tag{4.37}$$

Since λ, m, l have dimensions MLT^{-2}, M, L respectively, the dimension of the constant n is $(MLT^{-2}M^{-1}L^{-1})^{1/2}$, that is T^{-1}. Thus n has the dimension of a *frequency* or an angular speed.

Notice first that (4.37) is satisfied for all t if x has the constant value x_0, where

$$x_0 = \frac{g}{n^2} = \frac{mgl}{\lambda}. \tag{4.38}$$

For, if x is constant for all t, then $\dot{x}$ and $\ddot{x}$ are zero for all t. Hence the particle can be in *equilibrium* at $x = x_0$ with the restoring force exactly balancing the weight. However we can cause motion by disturbing the particle from its equilibrium position. For example, we could pull P downwards from the position $x = x_0$ and then release it from rest, or we could give Q a vertical blow. However the motion is caused, we have to use (4.37) to study how the motion develops.

Obviously the position $x = x_0$ has special significance. It is sensible therefore to use a new variable X defined by

$$X = x - x_0, \quad x = x_0 + X, \tag{4.39}$$

so that X is the displacement of the particle measured from its equilibrium position. Substitution of (4.39) in (4.37) gives

$$\ddot{X} + n^2 X = 0. \tag{4.40}$$

We can verify easily that (4.40) is satisfied if $X = \sin nt$ for then $\dot{X} = n \cos nt$ and $\ddot{X} = -n^2 \sin nt = -n^2 X$. Similarly it is satisfied if $X = \cos nt$. It is then easy to see that (4.40) is also satisfied by

$$X = \alpha \sin nt + \beta \cos nt, \tag{4.41}$$

for all values of the constants α and β. In fact we can show, but the proof is beyond the scope of this book, that every solution of (4.40) is of the form (4.41), that is (4.41) is the *general solution* of (4.40). We can also write (4.41) in the form

$$X = A \sin (nt + \phi), \tag{4.42}$$

where the constants α, β are related to the constants A, ϕ by the equations $A \cos \phi = \alpha$, $A \sin \phi = \beta$. Thus $A = (\alpha^2 + \beta^2)^{1/2}$. By combining (4.41) or (4.42) with (4.39) we find that

$$x = x_0 + \alpha \sin nt + \beta \cos nt = x_0 + A \sin (nt + \phi). \tag{4.43}$$

The motion of the particle is an *oscillation* about the equilibrium position, or centre, $x = x_0$. The oscillation is *periodic* in time, since the periods of the sine and cosine functions are both 2π. The value of x at any time t will be the same at time $t + T$, where

$$T = 2\pi/n; \tag{4.44}$$

T is the *period* of the motion. The maximum distance of the particle from its centre is A; this maximum distance is known as the *amplitude* of the motion. Motion satisfying (4.43) is called *simple harmonic motion*, usually abbreviated to *S.H.M.*, and it is extremely important because it occurs in very many different situations.

The values of the constants in (4.43) are determined by the way in which the motion is set up. This is illustrated by the next two examples.

Example 11 A particle of mass 2 kg is fastened to one end of a light elastic string, the other end of which is fastened to a fixed support. When the particle is hanging at rest vertically below the support the string is extended by 0·08 m. The particle is then pulled down another 0·04 m and released from rest. Find the period of the motion, taking g as 9·8 ms^{-2}.

We are given that $x_0 = 0{\cdot}08$ m and $g = 9{\cdot}8$ ms^{-2}.

$$(4.38) \quad \Rightarrow \quad n^2 = 122{\cdot}5 \text{ s}^{-2} \quad \Rightarrow \quad n \approx 11{\cdot}07 \text{ s}^{-1}$$

$$(4.44) \quad \Rightarrow \quad T = 2\pi/n \approx 0{\cdot}57 \text{ s}.$$

Example 12 For the motion described in Example 11, find

(i) the maximum speed of the particle;

(ii) the extension of the string 10 s after release.

(i) Equation (4.43) gives $\qquad x = x_0 + \alpha \sin nt + \beta \cos nt.$

When $t = 0$, $x = 0{\cdot}12$ m, $\dot{x} = 0$ ms^{-1}. Also $x_0 = 0{\cdot}08$ m.

$$\Rightarrow \quad 0{\cdot}12 \text{ m} = 0{\cdot}08 \text{ m} + \beta, \quad 0 = \alpha n$$

$$\Rightarrow \quad x = 0{\cdot}04[2 + \cos nt] \text{ m}. \qquad (4.45)$$

Note first that the minimum value of x occurs when $\cos nt = -1$, and this minimum value is 0·04 m. Since this is positive, the string is stretched throughout the motion, and we are justified in using (4.45) for all t. By differentiating (4.45) we can find $\dot{x}$. Using $n \approx 11{\cdot}07$ s^{-1} from Example 11, the result is

$$\dot{x} \approx -0{\cdot}44 \sin nt \text{ ms}^{-1}.$$

The maximum value of $|\dot{x}|$, the speed of the particle, is therefore about 0·44 ms^{-1}.

(ii) Putting $n \approx 11{\cdot}07$ s^{-1} and $t = 10$ s in (4.45) gives the extension after 10 s as

$$0{\cdot}04[2 + \cos 110{\cdot}7] \text{ m} \approx 0{\cdot}05 \text{ m}$$

(Note that in cos 110·7, the units of 110·7 are radians.)

Example 13 A horizontal platform is moving vertically so that its displacement x from a fixed horizontal plane at time t satisfies

$$x = a \sin^2 \omega t,$$

where a and ω are positive constants. Prove that the platform is in simple harmonic motion with its centre at $x = (a/2)$. Find the amplitude and period of this motion.

A particle is placed on the platform when $x = 0$ and is observed to leave the platform, travelling vertically upwards, when $x = (5a/6)$. Show that

$$4a\omega^2 = 3g.$$

We have to show that $x = a\sin^2 \omega t$ is the same as the expression for x in (4.43) for specific values of x_0, n, α, β.

$$\cos 2\omega t \equiv 1 - 2\sin^2 \omega t \quad \Rightarrow \quad x = (a/2)(1 - \cos 2\omega t).$$

This is (4.43) with $x_0 = a/2$, $n = 2\omega$, $\alpha = 0$, $\beta = -a/2$. Therefore the motion is S.H.M. with centre $x = x_0 = (a/2)$.

From (4.44) $$T = 2\pi/n = \pi/\omega.$$

The amplitude, the maximum value of $|x - x_0|$, is $(a/2)$.

Let the mass of the particle be m. The platform exerts a force on the particle; let its component vertically upwards be N. By Newton's Second Law

$$N - mg = m\ddot{x} = 2ma\omega^2 \cos 2\omega t$$

$$\Rightarrow \quad N = m(g + 2a\omega^2 \cos 2\omega t)$$

$$\Rightarrow \quad N = m(g + 2a\omega^2 - 4x\omega^2).$$

Clearly the value of N must be positive; the platform cannot give the particle a downwards push. Hence the particle leaves the platform when $N = 0$, which occurs when $x = 5a/6$.

$$\Rightarrow \quad g + 2a\omega^2 - 4(5a/6)\omega^2 = 0$$

$$\Rightarrow \quad 4a\omega^2 = 3g.$$

Exercise 4

1 A particle of mass 3 kg starts with velocity $(\mathbf{i} + 2\mathbf{j} + 3\mathbf{k})$ ms^{-1} from a point A with position vector $(4\mathbf{i} + 3\mathbf{j} + 2\mathbf{k})$ m. The particle moves under the action of a constant force $\mathbf{F} = (3\mathbf{i} + 4\mathbf{j} + 8\mathbf{k})$ N, and travels from A to a point B in 4 s. Find the position vector of B, and the speed of the particle when it reaches B.

2 A boy throws a ball vertically upwards by the side of his house. The ball leaves his hand at a height of 1·5 m above the ground. His sister is looking out of a window so that her eyes are 3·8 m above the ground. Using a stop watch, she times the interval between the ball going past her eyes upwards and going past them downwards as 1·3 s. Taking g as 9·8 ms^{-2} and neglecting air resistance, show that the speed of projection is 9·3 ms^{-1} correct to 2 significant figures. Find the speed at which the ball first hits the ground correct to 2 significant figures.

3 A man standing on a platform throws a ball vertically upwards with speed U. Immediately after the ball leaves his hand, the man and the platform descend vertically with constant speed V. Show that the time that elapses before the ball returns to the man's hand is $2(U + V)/g$.

4 A man drops a coin from a height of 2 m above the floor of a lift. Calculate, to 2 decimal places, the times taken for the coin to hit the floor of the lift in the cases: (a) when the lift is at rest; (b) when the lift is descending with constant acceleration 1 ms^{-2}. Take g as 9·8 ms^{-2}.

5 A golf ball, driven from a point P with an initial speed of 50 ms^{-1}, first strikes the ground at a point Q on the same horizontal level as P and 200 m from P. Neglecting air resistance and taking g as 10 ms^{-2}, find, correct to the nearest degree, each of the

two possible angles of projection and, correct to the nearest tenth of a second, the difference between the two possible times of flight.

Show that in the lower of the possible trajectories the ball could not clear a tree 30 m high anywhere in its path.

6 Two points A and B are situated on level ground at a distance a apart. At a given instant a balloon is released from rest at A, and simultaneously a bullet is fired from a gun at B. The balloon rises vertically with a constant acceleration f, and the bullet has an initial speed U at an angle α above the horizontal. Given that the balloon is hit by the bullet, show that $U^2 \sin 2\alpha = a(f + g)$.

7 A projectile has initial speed U at an elevation 2α above the horizontal. Its point of projection is at the foot of a plane inclined at an angle $\alpha(< \frac{1}{4}\pi)$ above the horizontal. The motion takes place in the vertical plane through the line of greatest slope of the plane. Given that the projectile strikes the plane at right angles, find the value of $\tan \alpha$.

8 An aircraft is flying with constant speed V in a direction inclined at an angle α above the horizontal. When the aircraft is at height h a bomb is dropped. Show that the horizontal distance R, measured from the point vertically below the point at which the bomb is dropped to the point where it hits the ground, satisfies

$$gR = \tfrac{1}{2}V^2 \sin 2\alpha + V \cos \alpha \sqrt{(2gh + V^2 \sin^2 \alpha)}.$$

9 A ball thrown with initial speed $\sqrt{(2gh)}$ strikes a vertical wall which stands at a distance d from the point of projection. Show that the point on the wall that is hit by the ball cannot be at a height greater than $(4h^2 - d^2)/(4h)$ above the point of projection.

Show also that the region of the wall within range of the ball is bounded by a parabola.

10 Water is thrown off the tyres of a racing car travelling at speed V along a wet horizontal circuit. The radius of the wheels (including the tyres) is r, and there are no mudguards. Given that there is no slipping between the tyres and the road and that air resistance can be neglected, show that, when $gr/V^2 < 1$, water reaching the greatest height above the ground comes from points on the tyre circumferences such that the radius through each of these points makes an angle θ above the horizontal, where $\sin \theta = gr/V^2$. Determine the greatest height in this case. Discuss what happens when $gr/V^2 > 1$.

11 A particle moves in a straight line on a smooth horizontal plane against a resistance kv per unit mass, where v is the speed of the particle. Obtain the equation relating $\dot{v}$ and v. Verify that this equation is satisfied by $v = Ve^{-kt}$, where V is a constant equal to the speed of the particle when $t = 0$.

12 A ball of mass m is dropped from a height H and falls. It is acted on by gravity, and an air resistance of magnitude mkv^2 where k is a positive constant and v is the speed of the particle when it has fallen through a distance x. Show that

$$v\frac{dv}{dx} = g - kv^2.$$

Verify that this equation is satisfied if $v^2 = (g/k)(1 - e^{-2kx})$. Deduce the value of v when the ball hits the ground.

13 A particle is released above the ground and, while falling, suffers a resistance kv^2 per unit mass, where v is its speed. Show that, however long the particle takes to hit the ground, v can never be greater than V, where $V = \sqrt{(g/k)}$. Prove that, when the particle has been falling for a time t, the value of v is $V(e^{2kVt} - 1)/(e^{2kVt} + 1)$.

14 A particle of mass m is projected vertically upwards with speed $U \tan \alpha$ in a medium which exerts on the particle a resistance of magnitude mgv^2/U^2, where v is the speed of the particle. Show that the greatest height attained by the particle is

$$(U^2/g)\ln(\sec\alpha).$$

Show also that the particle returns to the point of projection with a speed $U\sin\alpha$.

15 A ball of mass m is thrown vertically upwards with initial speed u. When the speed of the ball is v, the air resistance is kv^2, where k is a constant. Find the dimension of k, and hence show that (ku^2/mg) is without dimension. The particle returns to the point of projection after a time T. Deduce that there is a function F (which is not to be determined) such that $T = (u/g)\mathrm{F}(ku^2/mg)$.

16 A particle moves along a line of greatest slope of a rough plane inclined at an angle β to the horizontal, where $\tan\beta = (3/4)$. The particle passes through a point A when moving upwards with speed u, comes momentarily to rest at a point C, and subsequently passes through a point B when moving downwards at the same speed u. Given that the coefficient of friction between the particle and the plane is (1/4), find AC, and show that $AB = AC$.

17 A particle of mass 0·5 kg is moving at constant speed on a rough horizontal table in the direction of the constant unit vector **i**. The particle is acted on by three forces, namely (i) its weight, (ii) an applied force **P**, and (iii) the reaction of the table **R**. Given than $\mathbf{P} = (3\mathbf{i} + 8\mathbf{j})$ N, where **j** is a unit vector vertically downwards, and taking g as 9·8 ms^{-2}, determine **R** in the form $\mathbf{R} = (k\mathbf{i} + l\mathbf{j})$ N, where k and l are numbers. Hence determine the coefficient of friction between the table and the particle to 2 decimal places.

18 A square $ABCD$, of side 2 m, is fixed with the vertex A on a horizontal table and AC vertical. A particle starts from rest at C and slides down CB which is rough with coefficient of friction $(2 - \sqrt{2})/2$. Taking g as 10 ms^{-2}, show that the particle leaves the square with a speed $2\sqrt{5}$ ms^{-1}.

Further, show that the particle strikes the table about 2·4 m from A.

19 A light inextensible string hangs over a smooth, fixed pulley. To one end of the string is attached a particle A of mass $3m$ and to the other end is attached a scale-pan of mass $2m$ containing a particle B of mass $5m$. The system is released from rest with the hanging parts of the string taut and vertical. Find the magnitudes of the tension in the string and of the force exerted by B on the scale-pan.

20 A particle A of mass $2m$ is initially at rest on a smooth plane inclined at an angle α to the horizontal. It is supported by a light inextensible string which passes over a smooth, light pulley at the top edge of the plane. The other end of the string supports a particle B of mass m which hangs freely. Given that there is no motion, find α.

A further particle of mass m is now attached to B and the system is released. Find the acceleration of B in the ensuing motion.

21 Two particles A and B, of mass $3m$ and m respectively, are connected by a light inextensible string which passes over a fixed, light, smooth pulley. The system is released from rest with A and B at the same level and with the string taut. After a time t_0 the string breaks and it is sufficiently long for B not to hit the pulley subsequently. Show that, when B reaches its highest point, the vertical distance between A and B is gt_0^2.

22 A lift which, when empty, has mass 1000 kg is carrying a man of mass 80 kg. The lift is descending with a downward acceleration of 1 ms^{-2}. Taking g as 9·8 ms^{-2} and ignoring friction, calculate the magnitudes of the tension in the lift cable and of the vertical force exerted on the man by the floor of the lift.

The lift is designed so that during any journey the magnitude of its acceleration reaches, but does not exceed, 1 ms^{-2}. Safety regulations do not allow the lift cable to bear a tension of magnitude greater than 20 000 N. Making reasonable assumptions and showing your working, suggest the number of persons that the lift should be licensed to carry. [Hint: the magnitude of the tension in the lift cable is greatest when the lift is accelerating upwards.]

23 A particle of mass m is connected by an inextensible light string of length l to a fixed point on a smooth horizontal table. The string breaks when subject to a tension the magnitude of which exceeds mg. Find the maximum number of revolutions per second that the particle can make without breaking the string.

24 A particle of mass m is attached to one end Q of an elastic string PQ of modulus $3mg$ and natural length l. If the string will break when its total length is greater than $3l$, find the maximum constant angular speed of the particle when describing a horizontal circle with the end P of the string fixed.

25 A particle executes simple harmonic motion with centre O, the period of the motion being 3 s. Find the time taken by the particle to travel directly from instantaneous rest at a point A to the mid-point of OA.

26 A particle of mass m is at rest suspended from a fixed point by a light, elastic spring of modulus $2mg$ and natural length l. Find the period and the amplitude of the oscillations performed if the particle is projected vertically from this position with speed $\sqrt{(gl/2)}$.

27 A particle of mass m is attached to the middle point of a light, elastic string of natural length a and modulus mg. The ends of the string are attached to two fixed points A and B, where A is a distance $2a$ vertically above B. Prove that the particle can rest in equilibrium at a depth $5a/4$ below A, and find the period of oscillation if it is displaced slightly in the vertical direction.

28 One end O of a light, elastic string OA, of natural length $4a$ and modulus $3mg$, is attached to a fixed point on a smooth horizontal table. A particle of mass m is attached to the other end A. The particle is pulled along the table until it is a distance $5a$ from O and it is then released from rest. Show that the particle first reaches O after a time $(\pi + 8)\sqrt{(a/3g)}$.

29 The end A of a light, elastic string AB, of natural length 0·5 m, is fixed. When a particle of mass 0·3 kg is attached to the string at B and hangs freely under gravity, the extension of the string in the equilibrium position is 0·075 m. Calculate the modulus of elasticity of the string taking g as 10 ms^{-2}.

The particle is now pulled down vertically a short distance and released from rest. Show that the time that elapses before the particle first passes through the equilibrium position is about 0·136 s.

30 A particle of mass m lies on a smooth horizontal table and is attached to one end of a spring, the other end of which is fixed to a point on the table. The unstretched length of the spring is d and its stiffness is k. Show that the dimension of k is MT^{-2}. Assuming that the period T of oscillations of the particle on the table is given by a formula of the type

$$T = \lambda m^{\alpha} d^{\beta} k^{\gamma},$$

where λ is a dimensionless constant, find the values of α, β, γ for dimensional consistency.

Answers

Exercise 1

1 645°, $\sin\theta \approx -0{\cdot}97$, $\cos\theta \approx 0{\cdot}25$, $\tan\theta \approx -3{\cdot}85$
2 (a) $9{\cdot}46 \times 10^{15}$ m;
(b) $9{\cdot}46 \times 10^{12}$ km;
(c) $5{\cdot}88 \times 10^{12}$ miles
3 $10{\cdot}3 \text{ ms}^{-1}$
4 Only $s = ut - \frac{1}{2}gt^2$ is consistent
5 $[v] = L^2T^{-1}$, $[k] = T^{-1}$, $[l] = LT^{-2}$

Exercise 2

2 $[a] = LT^{-2}$, $[k] = T^{-1}$, $U + [at/(1 + kt)]$
4 $\omega = 4 \text{ s}^{-1}$, $c = 0{\cdot}5$ m, $\alpha \approx 1{\cdot}24$
5 $(x/X) \approx 1{\cdot}62$
6 Trap. rule: $s \approx 0{\cdot}25$ m; $g \approx 9 \text{ ms}^{-2}$, $k \approx 11 \text{ s}^{-1}$. Integration: $s \approx 0{\cdot}25$ m.
7 $v = gT\exp[x/(gT^2)]$
9 $\mathbf{r} = [a\mathbf{i}\cos\omega t + b\mathbf{j}\sin\omega t]$
10 $\dot{\mathbf{r}} = c\omega(\mathbf{i}\cos\omega t - \mathbf{j}\sin\omega t) + U\mathbf{k}$, $\ddot{\mathbf{r}} = -c\omega^2(\mathbf{i}\sin\omega t + \mathbf{j}\cos\omega t)$
11 $\Omega\sqrt{(\Omega^2r^2 + 4V^2)}$
12 $(\mathbf{i} - 4\mathbf{j}) \text{ ms}^{-1}$; $(11\mathbf{i} + 3\mathbf{j})$ m
14 50 m
16 (i) $u = 120 \text{ ms}^{-1}$, $t = 20$ s
17 (i) $d = b - Ut + \frac{1}{2}ft^2$,
(ii) $d = b + (V - U)t - V^2/(2f)$;
$d_{\min} = b - U^2/f$; when $U^2 < 2bf$ and $U > V$ the car catches the bus after a time greater than V/f
18 $\frac{1}{3}V\sqrt{3}$ (towards wall),
$\frac{4}{9}(V^2/a)\sqrt{3}$ (towards wall)
19 $\sin^{-1}\sqrt{(1 - \lambda^2)}$E of N, $d\sqrt{(1 - \lambda^2)}$, $d\lambda/[U\sqrt{(1 - \lambda^2)}]$

Exercise 3

1 $0{\cdot}8$ N
2 $\mathbf{v} = [(k\mathbf{V} + \mathbf{g})e^{-kt} - \mathbf{g}]/k$,
$\mathbf{a} = -(k\mathbf{V} + \mathbf{g})e^{-kt}$; Units: $k - \text{s}^{-1}$; $\mathbf{V} - \text{ms}^{-1}$.
3 $m = p = 2, n = 1$
4 $19{\cdot}3 \times 10^4 \text{ Nm}^{-2}$
5 $1{\cdot}677 \times 10^{-25}$ kg,
$5{\cdot}006 \times 10^{-17} \text{ kg ms}^{-1}$

Exercise 4

1 $(16\mathbf{i} + \frac{47}{3}\mathbf{j} + \frac{79}{3}\mathbf{k})$ m,
$\frac{1}{3}\sqrt{2390} \text{ ms}^{-1} \approx 16{\cdot}3 \text{ ms}^{-1}$
2 11 ms^{-1}
4 (a) $0{\cdot}645$ s, (b) $0{\cdot}67$ s
5 27°, 63°; $4{\cdot}5$ s
7 $\frac{1}{2}\sqrt{2}$
10 $r + (V^2/2g) + (gr^2/2V^2)$; $gr/V^2 > 1$ maximum height is $2r$.
11 $\dot{v} + kv = 0$
12 $\sqrt{[(g/k)(1 - e^{-2kH})]}$
15 $[k] = ML^{-1}$
16 $AC = (5u^2/8g)$
17 $k = -3$, $l = -12{\cdot}9$; $\mu = 0{\cdot}23$
19 $(21/5)mg$, $3mg$
20 $\alpha = (\pi/6)$, $\frac{1}{3}g$
22 9504 N, 704 N; assuming men of average mass less than 80 kg, licensed to carry no more than 10 persons.
23 $(1/2\pi)\sqrt{(g/l)}$
24 $\sqrt{(2g/l)}$
25 $0{\cdot}5$ s
26 $\frac{1}{2}l$, $\pi\sqrt{(2l/g)}$
27 $\pi\sqrt{(a/g)}$
29 20 N
30 $\alpha = \frac{1}{2}$, $\beta = 0$, $\gamma = -\frac{1}{2}$

Index